The Ostrich Factor

The Ostrich Factor

Our Population Myopia

Garrett Hardin

New York   Oxford
Oxford University Press
1999

Oxford University Press

Oxford   New York

Athens   Auckland   Bangkok   Bogotá   Buenos Aires   Calcutta
Cape Town   Chennai   Dar es Salaam   Delhi   Florence   Hong Kong   Istanbul
Karachi   Kuala Lumpur   Madrid   Melbourne   Mexico City   Mumbai
Nairobi   Paris   São Paulo   Singapore   Taipei   Tokyo   Toronto   Warsaw

and associated companies in

Berlin   Ibadan

Copyright © 1998 by Oxford University Press, Inc.

Published by Oxford University Press, Inc.
198 Madison Avenue, New York, New York 10016

Oxford is a registered trademark of Oxford University Press

Library of Congress Cataloging-in-Publication Data
Hardin, Garrett James, 1915–
    The ostrich factor : our population myopia / by Garrett Hardin.
    p.   cm.
    ISBN 0-19-512274-7 (alk. paper)
    1. Population--Environmental aspects.   2. Quality of life.
    I. Title.
    HB849.415.H37   1998
    304.2–DC21   97-39272

1 3 5 7 9 8 6 4 2

Printed in the United States of America
on acid-free paper

Contents

Acknowledgments

It is a pleasure to express my gratitude to the activist founda-
tions that have supported this work—namely, the Federation for
Immigration Reform, the Laurel Foundation, the Pioneer Fund,
Population Communication, Inc., and the Caroline H. Robinson
Social Service Fund—along with my heartfelt thanks to their
directors.

In completing this work, I have been much aided by my
research assistants, Tom Andres and Tracy Fernandez. Valuable
criticism has been contributed by my wife, Jane, and our sons,
Peter and David.

I express my gratitude for the unwavering support of the
administrators of my home institution, the University of California at Santa Barbara. A basic function of an academic institution
is the support of its "oddballs"—call them apostates, noncon-
formists, dissenters, iconoclasts, or what you will. In developing
a true "community of scholars" academic investigators must
guard against becoming what David Riesman calls "other-directed" agents. Those who study population questions find that the
principal *others* are those who stand to gain—temporarily at
least—from population growth. Merchants, real estate agents,

and mediamasters supported by advertising can exert considerable pressure on educational institutions. It is my hope that the present work will convince its readers that real benefits are to be gained by ignoring the world's ostriches as we take a long view of population questions.

The Pursuit
of Objectivity

We know that the tale of the ostrich that buries its head in the sand is mythical, a myth that is both ancient and confusing. In the 1st century A.D., Pliny the Elder said that the stupid ostrich thrusts its head and neck into a bush, imagining "that the whole of the body is concealed."[1] Not until the 14th century was sand substituted for the bush: the altered version endured to the present (though only as a myth, not as natural history). What accounts for the endurance of such a silly story?

A plausible psychological explanation can be suggested. Bearing in mind Wordsworth's insight that "the child is father of the man," we note that the varied responses of a small baby to a threatening and oh-too-close face fall into two classes. First, it may try to escape knowledge of the stimulus by burying its head in a blanket or in its mother's comforting bosom. (Some dictionaries call this behavior *ostrichism.*) Though we never really know the thoughts of another being (baby or not), it is reasonable to assume that the infant's mind moves along the following sort of logical path: "My world is what I see. If I do not see something, it does not exist. I will cause this fearful object to cease to exist by wiping out its image." In brief, the baby unknowingly resorts to Freudian denial. When a whole culture responds in this way, it is said to be in the grip of a *taboo*, to use a term brought from the South Seas by Captain James Cook in 1777.

The second explanation of a baby's behavior rests on an axiom advanced by Aristotle: "Man by his nature desires to know." This should be amended to "know and/or control." ("Practical" people are often satisfied with mere control.) A baby, by screaming, may cause help to appear. The infant's resort to this behavior bespeaks a primitive sort of knowledge.

Adults who indulge in ostrichism can be said to be observing a taboo, which closes off the search for causes. The taboo now laid on the subject of human population growth is far from total, but it does inhibit the search for causes. Thomas Robert Malthus brought the subject into the open in 1798,[2] and for a good half century it was a popular topic of public discussions. Then the dialogue slowly degenerated until, during the second half of the 20th century, population was considered a slightly laughable topic among many academics. Was this because human populations were no longer growing? By no means. In Malthus's day the world population was about 1,000 million; now it is nearly six times as great—but the topic is no longer very popular.

Malthus was an economist, but many of today's economists say there is no such thing as a population problem. The deniers maintain that the more people there are in the world, the more rapidly civilization will advance because there will be more Einsteins and Shakespeares to solve humanity's problems. It is worth pointing out that England today has 13 times as many people as it did in Shakespeare's time. "And where," we might ask, "are the 13 Shakespeares?" The world's 6 billion people should be more than enough to furnish whatever talents civilization requires. Evidently it takes more than mere numbers to produce a sufficiency of geniuses.

Ask yourself this question: what features of your daily life do you expect to be *improved* by a further increase in population? Will commuting time to work be decreased? Will streets and highways be less crowded? Will the air be cleaner? Will it cost less to get sparkling water to drink? Will vacation spots be easier to get to and less crowded when you get there? Will the extinction of interesting and valuable animals and plants come to an end? Will crime in the streets diminish? Will international conflicts taper off? There seems to be no end to the negative effects that can reasonably be expected from a further increase in population. At the present rate of population growth, it is difficult to be optimistic about the future; yet more than a few academic ostriches, their heads in the sand, continue to chant: "*We* see no population problems ahead."

▲ ▼ ▲ ▼ ▲

How does knowledge advance? In the natural sciences (e.g., physics, chemistry, biology), technological inventions are critical. The microscope, the telescope, and X-ray machines have brought much progress during the past three centuries. But when we look at the social scene, improvements are less obvious. Are there social inventions that are comparable to the technological ones of the natural sciences?

I believe there are; I will cite one instance. Early in Captain Cook's century, a systematized approach was devised to undermine the power of taboos (which were effective even before they were officially named). Living in a world of tight totalitarian controls, the Baron de Montesquieu showed, in his *Persian Letters* (1721), how gentle ridicule could undermine the power of taboos. His book purported to be a collection of letters written home by two Persian visitors to France. They commented at length on how strange the natives were. They began by expressing wonder that French men wore wigs, and they were astonished that French women never donned the pantaloons worn by proper Persian women. They called attention to the fact that, in the evaluation of money, the French people did whatever their number one magician demanded: if he said that one coin was now worth two, that was that. The French people also deferred to a magician in Rome who said that three people could sometimes be but one, while the bread and wine ingested in certain places on certain days were no longer bread and wine but something else (which was not well defined). What might have been condemned as dangerously seditious observations were no doubt made more acceptable by salacious accounts of sexual behavior in Persian harems.

It is important to note that the baron's book was not published in either Paris or Rome. The title page gave the place of publication as Cologne (Germany), and the publishing house listed was fictional. No author was given. The place of publication was in fact Amsterdam, a hotbed of free speech in 18th-century Europe.

In attaining the objectivity that is so desirable in the social sciences, this pioneer effort fell short of perfection: it ended up by being a nominal comparison of two cultures rather than an evaluation of one from a point of view untainted by unconscious assumptions. (Other writers after Montesquieu did not escape this shortcoming by substituting Turkish and Chinese observers for Persian.) Nevertheless, in getting people to think about taboo subjects, Montesquieu made a social invention comparable to the microscope and telescope of the physical sciences. It is true that fashioning mental machinery is not as easy as rearranging material objects like nuts and bolts, but over the centuries the most creative social thinkers have had some success in consciously using fictions such as foreign visitors to free their minds of fashionable cant.

In 1759, Adam Smith (before he became an economist) put forward a less dramatic version of this sort of invention when, in *The Theory of Moral Sentiments*, he postulated an "Impartial Spectator" as a hook on which to hang his social insights. In the 19th century a more colorful concept, *the Man from Mars*,[3] became the gimmick of choice. Since they are drawn from no known culture, Martians can be presumed to achieve great objectivity, uncontaminated as they are by earthly assumptions.

Objectivity is particularly needed when investigators take up the problems associated with the size of human populations. The Man from Mars would surely ask, "Why don't you try to prevent further increases in population? Or even try to decrease the present overpopulation—by acceptable means?"

The heart of the difficulty lies in the phrase *by acceptable means*. If we already knew a means that we could all agree was acceptable, we could install a system of population control right now. Conventional ethical principles often prevent us from even looking at proposals that might do the job. Odd. No one expects the physics of 50 B.C. to tell us how to launch a spaceship. But apparently many people are sure that the 2,000-year-old ethics developed in Near Eastern villages is all we need to solve all the moral problems created by our cleverness in applying the natural sciences to a world community that is measured in the billions. The rest of this book attempts to achieve an objective Man-from-Mars comparison of competing ethical assumptions.

Many scholars now recognize that the disciplines of economics, ecology, and ethics share a common problem, namely: *to discriminate among limitless demands in a world of limited resources.* Some contemporary economists reject this generalization because the economics that became orthodox in the two centuries after Adam Smith built its theories on the unstated belief that limits do not exist; or, if limits do exist, they must not be allowed to curb growth. Perpetual growth has become a secular religion built on the assumption that *growth = progress.*

Fortunately, the hybrid discipline of ecological economics has now been born. Limits are incorporated into the very foundation of its revolutionary structure. Much of the old economics is now regarded as myth. Mythic priests still hold the high ground, of course, but their days of dominance are surely numbered.[4]

Traditional ethics also often fails to take account of the inescapable limits of the world. Ecology, however, has—in a good sense—been a *limited* science from the very beginning. The concept of a limited environmental *carrying capacity* (which many orthodox economists ridicule) plays a central role in ecological thinking. Ecologists admit that the universe may ultimately prove to be infinite in extent; but in the short term—the next 10 centuries, say—the earth, together with abundant solar energy and skimpy meteoritic dust raining in on it, will set practical limits to what *Homo sapiens* can do.

One of the tasks of this book is to show how ethics and economics are transformed by paying attention to the insights of ecology. The power figures of contemporary society—journalists and politicians—see their interests served best by denying the reality of limits, thus turning the topic of population into a virtual taboo.

Disputes over population started in earnest with Malthus. He was not the first to take population seriously; but, partly for accidental reasons, his writings were the first to lead to a sustained, if sometimes underground, discussion of the subject.

He never came up with a convincing proposal for avoiding overpopulation. His many successors have done little better.

I know this because I wrote one of the nonsolutions myself, namely, *Living within Limits*, published in 1993. A knowledgeable critic, Mark Sagoff, said that my book "provides little guidance as to how to achieve" the goal of preventing overpopulation.[5] He was right. Like others before me, I was inhibited by unacknowledged taboos against taking a Darwinian approach to population. Borrowing a term from equestrian sports, I *balked* before leaping the hurdle. At the risk of coming a cropper, I approach the Malthusian barrier once more.

The foundations of traditional ethics will be examined and ways suggested for modifying them to fit our new vision of the world. In evading the suppression achieved by time-honored taboos, I will present more questions than answers, but such is the nature of the Man-from-Mars approach. Taken seriously, the Martian path ultimately produces answers. The best description of this path has, I think, been given by Hans Spemann, who was awarded the Nobel Prize in 1935 for throwing much light on how an *apparently* simple fertilized egg can develop into an obviously complicated multicellular animal. The necessary attitude of the person who succeeds in elucidating the complexity that can grow out of simplicity was captured in Spemann's words, which bear repeating before we examine the *apparently* simple phenomenon of population growth:

> I should like to work like the archeologist who pieces together the fragments of a lovely thing which are left alone to him. As he proceeds, fragment by fragment, he is guided by the conviction that these fragments are part of the whole which, however, he does not yet know. He must be enough of an artist to recreate, as it were, the work of the master, but he dare not build according to his own ideas. Above all, he must keep holy the broken edges of the fragments; in that way only may he hope to fit new fragments into their proper place and thus ultimately achieve a true restoration of the master's creation. There may be other ways of proceeding, but this is the one I have chosen for myself.[6]

To "keep holy the broken edges" of knowledge—what a poetic expression of the humility the investigator should strive for in tackling a fiercely difficult problem! The path to understanding human populations is not a straight one: the "broken edges" we must examine include the behaviors of crickets, birds, and other nonhuman creatures. Like the Duke in Shakespeare's *As You Like It*, I believe that, under the pressure of adversity, the examined life will disclose "tongues in trees, books in the running brooks, / Sermons in stones, and good in everything." Well— almost everything.

Tertullian's Blessing

Mark Twain defined a classic as "a book that people praise and don't read." The treatment received by Malthus's *Essay on Population* during the past century suggests a somewhat perverse redefinition: "a book that people call *discredited* without bothering to read it." For more than a century Malthus's essay has been a discredited, oft-cited, and rarely read book.

What's wrong with Malthus's argument? In his second chapter he put forward two major theses: first, if unhindered, population "would increase in the ratio of—1, 2, 4, 8, . . . &c."; second, the means of subsistence increases only as the series "—1, 2, 3, 4, . . &c." Thus did he account for the fact everywhere observed that, in a finite environment, population growth ultimately comes to a halt.

Implicitly Malthus realized that the environment has a limited *carrying capacity* for living things, but this term had to await the coming of the 20th century. When one thinks in terms of total carrying capacity, the comparative rates of population growth and subsistence increase become almost irrelevant.

What about the number series themselves? The first is on absolutely sound ground; the second is unprovable nonsense.

Malthus called the first ratio geometrical; following the conventions of calculus, we now call it *exponential.* Money put out at perpetual compound interest is a well-known example: no matter how small the constant rate of compound interest may be (provided it's greater than zero), the sum of capital + accumulated interest will eventually reach beyond any definite sum you can name. Given enough time, $10 earning interest of only 0.0001 percent per year will eventually make you a millionaire. Likewise, a human population that is increasing at an average rate of 0.0001 percent per year will eventually overburden any finite world whatever. (Exactly when this will happen is, by comparison, a trivial matter.)

In trusting the two rates to prove his point, Malthus was betting on one winning and one losing horse. The mathematical reality of money at interest eventually overwhelming finite resources must have been realized as early as money lenders existed. But perhaps not many people extended this insight to problems of population growth. In an ever-threatening world, every species *must* be capable of increasing at some rate if it is to survive. In each population there are some genetic lines that reproduce more rapidly than others. As time passes, rapidly reproducing variants replace the slower ones. Mathematical realities have biological consequences.

The human implications of this progression were realized in the 3rd century A.D. by the Christian apologist Tertullian. Why, he asked, is the human population so vast that we are a burden to the earth, which can scarcely provide for our needs? (The world population in Tertullian's day was perhaps 150 million; it is now nearly 40 times as large.) In a short passage of *De Anima,* Tertullian explained the very real value of events that are customarily viewed with dismay.

> . . . [A]s our demands grow greater, our complaints against nature's inadequacy are heard by all. The scourges of pestilence, famine, wars, and earthquakes *have come to be regarded as a blessing* to overcrowded nations, since they serve to prune away the luxuriant growth of the human race.[1] (emphasis added)

Since Tertullian's time the Christian world has grown less tolerant of death. Encouraged by Scripture, Christians readily assume that God created the world for the pleasure of humanity. Not many Christians in the 20th century speak of pestilence, famines, and wars as *blessings* to our race. Tertullian, by using the phrase "have come to be regarded," implies that this was a common view in his day. Unlike modern Christians, Tertullian's contemporaries were probably not in the least shocked by what we may be tempted to call his heartless assertion—which will now be defended.

Notice how Tertullian defends the apparently harsh human and natural catastrophes by reminding us of their very real consequences, namely, the "pruning away" of "luxuriant" population growth that threatens to produce even greater suffering. Opponents of Tertullian try to brush aside the ethical problem by finding emotion-laden terms for what they want to reject. Such is the approach of *moralistic ethics*. In contrast, *consequential ethics* seeks to list all the reasonable alternatives, choosing from among them after comparing what appear to be the consequences of each.

There is a simple way of calling on the resources of consequential ethics: adopt the path used by Karl Jacobi in the 19th century whenever he was stuck in trying to solve a mathematical problem: *Invert, always invert.* Applied to the area of ethics this means that when the best solution we can find using conventional means is no more than semisatisfactory, *invert.* Suppose there were no such disasters as pestilences, famines, and wars: would suffering increase or decrease?

By favoring more rapid reproduction, nature favors the rapid exhaustion of resources in a limited world—which is the only kind of world we have a chance of inhabiting. When resources are exhausted, what does life become like? As far as food is concerned, starvation takes over when the daily supply of calories is less than the minimum required for living. As for all the other resources, for each of them life becomes less bearable as the per capita allotment is reduced. A Jacobian inversion would lead to a world more heartless than the one God and nature have furnished us. Q.E.D.: forces that act to curb the natural rate of increase should not be dismissed for their "heartlessness."

The Jacobian inversion now joins the mythical Man from

Mars as a major component in the armamentarium of ethics and the social sciences. Simple moralistic conclusions are replaced by consequential analysis.

Implicitly, Tertullian was thinking in terms of limits and carrying capacity. The paramount assumption of practical population theory (toward the expression of which both Tertullian and Malthus were struggling) can be added to an Ecological Decalogue:

Thou shalt not transgress the carrying capacity.

Tertullian and Malthus only implied this 11th Commandment. Civilization, if it is to survive, must someday frankly bow to its wisdom.

▲ ▼ ▲ ▼ ▲

I think that the discreditors of Malthus generally have in mind his arithmetic series: unlike the geometric series, no excuse can be given for its existence (nor does Malthus give one). Intuitively, we may feel that the progress of technology is less than the achievable increase in population, but how do we measure technological progress? By gross national product? By community happiness? By charting the number of patents? For a variety of reasons we reject all these possibilities. A great many disparate measures would somehow have to be combined; we don't know where to start. What relative weighting would we give to progress in computer technology compared with progress in perfecting a satisfactory electric car? Toward the end of the 20th century it was noted that the technical capacity of computers was doubling every 18 months. Contrast that with the rate at which we have worked toward the longed-for quiet, pollutionless electric automobile. On the latter, we've made some progress during this century, but it is doubtful if the doubling of this progress (however measured) has taken less than 20 years. If we survey the whole field of invention we become even more confused. There seems to be no hope of rescuing Malthus's arithmetic ratio.

In addition, Malthus is blamed for another area in which his critics inferred more than he implied: the predicting of future population growth. Malthus made no explicit numerical predictions that I know of, but the *flavor* of his dissertation is that of a person who thinks his world was very close to the end of its population growth. Though Malthus did not explicitly predict future populations, his rhetoric is such that it is fair to believe that he foresaw no significant increase in the carrying capacity of the earthly environment. In fact, however, world population grew about sixfold after Malthus published his essay.

At every moment, anyone who dares to predict the future depends largely on the projection of present trends: but as the microbiologist Rene Dubos has said, "Trend is not destiny." Malthus wisely never put much rhetorical force into his predictions of future population size. He deserves neither positive nor negative credit in this area.

A word about *prediction.* Embarrassing experiences, coupled with Dubos's warning, have led demographers to state repeatedly that they do not make predictions: only *projections*—projections of present trends. Trends may change with little warning. After two repetitions in the daily press, what begins as a projection metamorphoses into a prediction in the minds of readers. In spite of their warnings, demographers are repeatedly castigated for making predictions that don't come true.

▲ ▼ ▲ ▼ ▲

Malthus lived so near the beginning of the industrial-scientific revolution that he, like many others, did not suspect the rapid changes that were coming. To excuse historical short-sightedness, we note that the earlier period now called the Renaissance, though it began in the 14th century, was not given its name until the 19th. Malthus did not see that the earth's carrying capacity was being rapidly increased by human inventiveness. And it was only in the 20th century that the philosopher Alfred North Whitehead proclaimed that "The greatest invention of the nineteenth century was the invention of the method of invention." Dying in 1834, Malthus hardly had time to become aware that attainable subsistence was then beginning to grow exponentially.

To return to Tertullian, note (in the preceding quotation) the figure of speech "to prune away." This is eminently agricultural. Pruning is an action well known to farmers and orchardists—people bonded to the rural life. Growing things—*and getting rid of superfluous living material for the sake of a better harvest*—was once a familiar practice to many. But now, less than 2 percent of the American population lives directly off agriculture—1 person in 50. As far as urbanites are concerned, milk comes in waxy cartons; capons and steers are such great mysteries that they are seldom named. (And not 1 gourmand in 100 knows the origin of "Rocky Mountain oysters.")

Pruning is now an exotic action to most; pruning for the sake of a better future is, to many urbanites, unthinkable. Yet it is pruning, *or some unspoken but acceptable substitute*, that must be built into every workable population program to produce what we now call a *sustainable policy*.

The limitless view of human resources became the dominant view of economists and other social scientists in the 20th century. As late as 1977, the sociologist Daniel Bell said: "If one thinks only in physical terms, then it is likely that one does not need to worry about ever running out of resources."[2] Whether or not the majority of economists agree with this, it is significant that most current theories of economics are built on a hidden assumption of perpetual growth. Many economists assert that this *is* true; others assert that it *must* be true—an interesting criterion for determining what is true and what is not. For instance, the economist Wilfred Beckerman once said: "A failure to maintain economic growth means continued poverty, disease, squalor, degradation and slavery to soul-destroying toil for countless millions of the world's population." The joining of hardheaded businessmen aiming at profits with idealists yearning to do away with poverty has created a cozy conspiracy with powerful political force.

(The reader, whether conservative or liberal, may take offense at the word *conspiracy*. But let it be pointed out that, from its etymological roots, *con-spiracy* primarily means "a breathing together." It does not presume midnight gatherings of political activists. Without thinking much about the matter, conspirators act together because they are just breathing the same atmosphere of assumptions. Unspoken taboos are the glue that holds conspirators together.)

▲ ▼ ▲ ▼ ▲

In the second half of the 20th century, the comfy association of hardheaded businessmen and softhearted liberals was shocked by the sudden appearance of ecologists and environmentalists. More than all others, Rachel Carson's book *The Silent Spring* (1962) was the stimulus for a new intellectual revolution.

The new message was simple: *growth has its price.* Some things get better as growth continues; other things get worse. Because of fundamental mathematical and physical laws, the second effect eventually overwhelms the first. But just try to find any recognition of this "consequence of scale" in an economics text!

If you have access to a collection of elementary economics texts, look in their indexes for entries under "economies of scale" and "diseconomies of scale." The first category will be in almost every book, but very few books will have even a single entry in the second category. The *implied* moral in the unbalanced books is obvious: "we can't have too much growth." Taboo discourages us from taking a total view of the effects of size on the well-being of human populations. For this willful blindness, society ultimately pays a price.

There is a perilous gap between natural scientists and economists. Since the time of Epicurus in the 3rd century B.C., scientists have recognized the primacy of *conservative laws,* i.e., laws stating that the two sides of an equation must balance. What is gained on one side must be lost on the other. In chemical reactions, mass is conserved. Energy is also conserved. (Or, since Einstein, mass-energy is conserved because there is a conserving transformation of one entity into the other.)

Economists make a brave show of starting out an introductory college course with the conservative statement that "There's no such thing as a free lunch." But, acknowledging the power of profit-seeking businessmen and praise-seeking liberals, they soon slip in evasions of scientific conservatism, sporting such poorly defined fancies as *win-win* situations and *supply-side economics.* The future will no doubt bring new linguistic evasions of the truth. Some of these escape hatches are exposed in the next chapter.

How to Lie
with Learned Words

About half a century ago Darrell Huff, statistician, significantly advanced his profession when he published a little book called *How to Lie with Statistics*. It was an instant success, remaining in print for decades. An annotated display of bad examples in any field is often more educational than an adoring exhibition of the good. Is this because our inherent jealousy makes us pay closer attention to the errors of others than to their successes? Perhaps; but motives are less important than consequences. Effective education uses any tool that works.

Since the publication of Huff's book, a variety of authors have written at least seven other "How to Lie" books. These manuals tell the serious student how to lie with charts, maps, paradigms, cognition jargon, and physics. The implied purpose is, of course, to teach neophytes how to recognize falsehood in the writings of others while encouraging them to strive for the unvarnished truth in their own work. Let me join the crowd. I hope my examples will help arm the reader against unscrupulous word technicians.

▲ ▼ ▲ ▼ ▲

Population literature is replete with errors, not all of them innocent. Nonscientific questions get entangled with scientific ones. This point is well illustrated by a quotation from the 19th century published in the Supplement to the *Oxford English Dictionary* in 1972. The passage indicates a surprising reversal in the acceptance of certain medical procedures over a century's time:

> 1886 E. B. FOOTE *Radical Remedy in Social Science v. 89*
>
> Where it becomes a necessity to decide between lawful abortion and unlawful contraception, they [physicians] prefer to break the man-made law against contraceptics rather than the natural law against abortion.

The typical modern reader, naive about ancient distinctions, may wonder what the difference is between *law* and *natural law.* "Law" certainly applies to any statute passed by a legislature. What, then, is "natural law"? Investigators of the natural sciences (physics, chemistry, biology, etc.), since they are investigating the laws of nature, are likely to assume that what they are dealing with is called "natural law." But European history gives us quite a different meaning. Long before the efflorescence of science, men of the church claimed squatters' rights to the term natural law, by which they referred to the arrangements made by God, as interpreted by ecclesiastics, who almost invariably had no training in science. Where there is a conflict, churchmen maintain that their natural law preempts statute law. What men of the cloth assert to be natural law receives no validation from scientists.

What would our mythical Man from Mars write home about this? Probably something like the following: "The Earthling's 'natural law' is neither natural nor law. Its correctness is not proved; it is merely asserted. I hope in a later epistle to tell you what nonscientific Earthlings hope to achieve by adopting this deceptive language."

In the meantime the Martian observer asserts, with confidence, the following:

Earthly language serves two contradictory purposes: To facilitate thought and to prevent it.

Skill in the manipulation of words (whether written or spoken) has long been accepted as the mark of an educated man or woman. But no doubt every virtue has its absurd extreme. A famous example of ridiculous language was furnished by the English philosopher Herbert Spencer, who beat the drum for biological evolution several years before Darwin published his *Origin of Species*. Neither in his own time nor in ours has Spencer been given much credit for his originality in the field of biology. A casual examination of a famous statement of his shows why scientists have been willing to ignore this pioneer:

> Evolution is an integration of matter and concomitant dissipation of motion; during which the matter passes from an indefinite, incoherent homogeneity to a definite, coherent heterogeneity; and during which the retained motion undergoes a parallel transformation.[1]

The stunning clarity of this passage moved a contemporary mathematician to pen a parody: "Evolution is a change from a nohowish, untalkaboutable all-alikeness, to a somehowish and in-general-talkaboutable, not-all-alikeness, by continuous some-thingelsifications and sticktogetherations."

Now that that little matter has been cleared up, we are ready for a gibe that appeared in the English humor magazine *Punch* in the 1920s. As George Orwell retells the story, a young man was informing his aunt that he intended to become a writer. "And what are you going to write about, dear?" asks the aunt. "My dear aunt," says the youth in a statement that brought an end to the conversation, "One doesn't write about anything, one just *writes*."[2]

Conventional education in arithmetic and statistics is useful but not sufficient to deal with the sorts of problems that now rapidly approach us. As intellectual development proceeded over the centuries, it became ever clearer that the nourishment of good science required something more than glib literacy. Someone in the 1950s named this something *numeracy*, the meaning of which can best be conveyed by examples. Just counting is not enough; the routine work of statisticians, useful though it may be, does little more than touch on the domain of numeracy. Determining the numbers is only the first step in a numerate adventure. Next, one wants to know the *ratios* of interacting numbers, the *rates* at which they are changing, the *projections* of these rates, and their *competitive interaction*. Is the *extinction* of a competitive interactant on the horizon? What foreseeable extinctions are *significant*, and which ones are merely trivial in their consequences? All the italicized terms above are words, but answering the questions raised requires mathematical operations. The questions are numerate: they seldom are usefully dealt with by the merely literate.

The ethical questions raised by birth control are inherently numerate. It's no use trying to solve them with words like *sinful*, *unnatural*, *wicked*, and *unsanctified*. Having a baby is inherently neither good nor bad: it's a question of the numbers involved. Is it the first baby in the family or the sixth? Are the quantities—numbers again!—of the resources the developing child may draw upon either great or small? Is the foreseeable future of the environment supportively generous or dangerously diminishing? Numbers, numbers, numbers. . . .

Members of society who are wealthy, or whose livelihood is guaranteed by an institution, or whose religious commitments ensure that they will never have to choose between abortion (or other forms of birth control) and being saddled with the many burdens of parenthood—all these *sheltered classes* can easily approach all moral problems on an exclusively literate plane, with comforting words that give no hint of numerate realities, including the afflictions that time will bring. Margaret Sanger's experience as a nurse in daily contact with the wretchedly poor made her see the numerate realities that were effectively invisible to the sheltered classes—until she rubbed their noses in raw life. Opening the eyes of the socially blind required the creation of new terms: *birth control* in 1914 and *planned parenthood* in the 1930s. Literate approaches frequently deceive, but (with imagination) words can be made to serve the goals of intelligent numeracy. Compassionate souls soon see that all of society benefits when women are freed from the necessity of bearing *unwanted* babies. (It is remarkable how often a human ostrich who seeks to impose compulsory pregnancy and mandatory motherhood on women lightly belittles a woman's request for an abortion as being no more than a "whim.")

The substitution of *birth control* for *contraception* constituted a displacement of pedantic language. The later introduction of *family planning* led to better relations: *planning* always sounds good, and the word emphasized the importance of time and the future in reaching family decisions.

An additional benefit of the new approach soon became apparent. Family planning clinics not only promote contraception, they also offer gynecological services that help diminish infertility among women. One might say that these clinics play both sides of the street, thus gaining more support than would be accorded purely reductive practices.

It is a historical fact that the earliest supporters of birth control were people who also thought it was time for population control. Unfortunately, the two functions—birth control and population control—are frequently confused. Strictly speaking, birth control is a task of the individual woman (or married couple), whereas population control can only be achieved by group action. For instance, considering the present state of public health, the stability of a population can only be achieved by each couple's having 2.1 children. But how can a woman produce one-tenth of a child? There must be some sort of community decision to maintain any particular *average* number of children per family. Even professional demographers sometimes get confused. In 1986, in his address as retiring president of the Population Association of America, Paul Demeny felt obliged to begin with this sentence:

> The essence of the population problem, if there is a problem, is that individual decisions with respect to demographic acts do not add up to a recognized common good—that choices at the individual level are not congruent with the collective interest.[3]

A thoughtful review of the statement by Tertullian (see chapter 2) makes it clear that the Christian father perceived the opposition of these two interests—the individual's and the group's. Tertullian must have been fully aware of the undesirability of painful death *for the individual*; but he realized that whenever a community consists of too many people for the resources available to it, heavy mortality can then actually improve the conditions of life for the lucky survivors. So, as Demeny says, the interests of the community as a whole are not entirely congruent with the interests of its members considered simply as individuals. Among about 2 billion people now—the miserable members of desperately poor countries—the conditions of life are so poor that the heavy mortality of a great plague might indeed bring about a subsequent improvement for the reduced population.

A triple benefit is realized whenever a couple practices birth control:

First: to the parents, who are not forced to produce and raise a child when they are not ready for it;

Second: to the child who otherwise might be born, since it need not be subjected to the real risks of being raised by unwilling and possibly embittered parents;

Third: to the community as a whole, which (if it is already overpopulated) is likely to suffer more civic disorder whenever the proportion of unhappy citizens increases.

The person who contributes money to family planning organizations may have in mind the third benefit. But the awareness of these benefits may be slight compared with the blessing experienced by the woman who is spared a lifetime of more or less reluctant service to an unwanted child. Financial support directly promotes birth control. Its effectiveness in population control is debatable. This unsolved problem will be brought up again and again in the remaining chapters, in the firm belief that looking at a problem with the clear eyes of a mythical Martian is more productive in the long run than resorting to the behavior of a mythical ostrich. Even the most difficult problem is better seen clearly than denied totally.

Adopting this approach, however, is sure to lead to a doubting of commonly held beliefs, always a painful experience. Among the most important of the beliefs of our time are the ones that most people, in the unconscious hope of sparing themselves the ordeal of doubt, take to be *indubitable*. What light can we throw on beliefs such as these?

Foundations
of Activist Science
By Right or By Default ?

A rigorous discussion of the foundations of a subject is not everyone's cup of tea. However, following any later proposal to derive action recommendations from knowledge, one should anticipate political attacks. There must be a justification of the application of the more basic principles to the problems of life.

Investigating human motivations, Aristotle began his *Metaphysics* with the bald assertion that man, by his nature, desires to know. In practice, this impulse can lead the critical thinker into an *infinite regress* of logically related statements. For example, told at the outset that Z is the case, the critic asks:

"Why?"

"Because Y implies Z."

"Why is Y true?"

"Because X implies Y."

"And what about X?"

As any parent of an intellectually active 3-year-old knows, there is no graceful escape from such a potentially infinite series of questions. Adults who are made uneasy by this reality may meekly accept a "final cause" or some ultimate god vouched for by a religion. In practice, the abstract symbols X, Y, and Z used above often stand for a succession of statements like the following: "My mother told me; the priest told my mother; the bishop told our priest; the cardinal told the bishop; and the cardinal was told by the Pope in a solemn conclave held in Rome." The links in such a sacred chain of certitude are eminently *personal.*

By contrast, those whose adult life is a continuation of the questioning period of childhood have to become reconciled with the practical necessity of stopping a disturbing series of questions at some arbitrary point. To get the world's work done, philosophy, religion, and everyday life all have to resort to verbal formulas that stop discussion.

No doubt primeval religion evoked the earliest discussion stoppers. The licensed keeper of the stabilizing anchors of society was, for a long time, a priest. He—it usually was a he—was authority incarnate. When he intoned, "God says so," he was presumed to be reporting the decision of a higher Authority (invisible though It might be).

▲ ▼ ▲ ▼ ▲

Today's intellectual world was significantly shaped by the historical period we call the Enlightenment. A watershed of opinion came with the American and French Revolutions at the end of the 18th century. In the political sphere, the stabilization of social life that had been brought about earlier by religious authority was now increasingly supported by an assertion of universal human "rights": the right to life, the right to individual freedom, the right of free speech, and so on.

Having renounced the solace of holy authentication, philosophers attempted to bring human actions into harmony with the objective facts of science. During the Enlightenment the conflict between the older way and the newer grew intense, producing vigorous attacks on scientific ideals, principally by defenders of the Catholic Church. The defense continues to this day. In 1968, professor of philosophy Germain G. Grisez of Georgetown University, speaking of controversial reforms, said that if one is Catholic, one is necessarily a papist, and "one cannot say, 'Rome has spoken, but the cause goes on.' One has to say, 'Rome has spoken, the cause is finished.'"[1]

In what may have been an unconscious effort to escape the *tyranny of the personal*, reformers of the revolutionary period took up, and developed further, an idea that has ancient roots, namely, the idea of impersonal *rights*, which we may say (risking circular reasoning) all human beings are presumed to enjoy "as a matter of right." Explicitly legislated rights are presumed to be derived from universal rights, which may be only implicit. It is assumed that such rights derive from human nature, which is everywhere the same. Such rights are called *natural rights*.

In our day, it is almost a foregone conclusion that people will try to solve every controversial problem by calling on rights. But rights, as the demographer Paul Demeny has pointed out, "are almost empty of content. They can be given meaning and content only with reference to local conditions." And local conditions almost always bring scale into the picture.

For example, pedestrians, walking at a slow pace on an uncrowded sidewalk, can insist on their right to walk where they please: no great harm will be done to the community. By contrast, it is suicidal for the drivers of our multitudinous automobiles to insist on the right to drive on the left or the right, as suits their fancy. Local conditions, not abstract rights, are the decisive ethical factors. It is clear that the rhetoric of rights must yield to the reality of particular times and places. The most enthusiastic defenders of rights seem to be guided by this sort of view of the origin of human practices: preexisting rights ⟶ then legislated laws.

Jeremy Bentham (1748–1832) supported the reverse theory: legislated laws ⟶ then deduced rights. Bentham warned us against ascribing too much to the magical word *right*:

> Right is the child of law; from real laws come real rights, but from imaginary law, from "laws of nature," come imaginary rights. Natural rights is simple nonsense, natural and imprescriptable rights rhetorical nonsense, nonsense upon stilts.[2]

In reading this we have to be aware that Bentham is not using *laws of nature* to stand for what we now call scientific laws, such as the laws written in physics books. The *natural rights* he refers to are related to the natural law discussed in the previous chapter—laws asserted by and supported by men of religion. Such natural law is no concern of the sciences, which have come to prefer the word *principle*. Bentham was repelled by the use of ambiguous "laws" to establish doubtful rights.

It is interesting that the passage from Bentham quoted above was written in 1791 and first published—*in French*—in 1816. It was not published in English (Bentham's own language) until 1843—11 years after the death of the author. The delay leads one to smell a taboo. Even today Bentham's view is presented unfairly in some quarters. Evidently the ostrich still has its head in the sand.

A further indication of the immaturity of rights at the present time is this: in most cases, rights are claimed by spokespersons who *make no mention of responsibilities*. In the real world, the satisfying of a right is not cost-free. If I have a right to expensive medical care, who is responsible for its cost? The standard Marxist formula that is wheeled out whenever this issue is raised gives the appearance of matching responsibilities to rights—but it doesn't. "From each according to his abilities, to each according to his needs" sounds almost like a fair balancing of benefits against costs because a careless audition can lead the hearer to presume that the first "his" is the same as the second "his."

After correcting for this grammatical mistake we then ask: what if the quantities hidden beneath the word *according* are not equal? In a true welfare state, the nation is expected to pick up the hospital bill for the individual, no matter how big it is. When such a system is set in place it creates what is called a *moral hazard*—the felt need tends to escalate without assignable limit. Since ours is a limited world, an intolerable equation is thereby set in place. In any world that is limited in resources but not in demands, moral hazards require mandatory matching of rights and responsibilities. Humankind has solved many technical problems over the past generations; it has scarcely begun to solve the problem of socialized medicine. For the present, we can do little more than mention its existence.

▲ ▼ ▲ ▼ ▲

Every highly structured *personal* institution has to have its rhetorical anchors. The Russian czar had his ukases; the Vatican has its encyclicals, some of which are presumed to be infallible. What about science? Since it is impersonally focused, how should a potentially infinite regress of supporting statements be anchored? The chain of anchoring statements is quintessentially *impersonal*: each element of an impersonal experiment that needs to be justified finds its legitimation in an additional observation or experiment—which is also impersonal. Those who doubt any generalization in science can find its justification in the impersonal experiments they themselves can carry out and observe. The attitude of scientists was well expressed in the 17th century when the Royal Society of London took as its motto the words *Nullius in verba*. Loosely translated, this means "On no one's authority."[3] If the word *authority* must be used, scientists agree that "Only nature is our authority."

Martians agree; ostriches get sand in their eyes.

To see how changing from personal authority to impersonal authority changes the nature of truth, we cannot do better than go back 2,000 years to the time when Epicurus explained why the conservation of matter—its noncreation *and* its nondestruction—is the only reasonable assumption to make about the world:

> Nothing is created out of that which does not exist: for if
> it were, everything would be created out of everything
> with no need of seeds. And again, if that which disap-
> pears were destroyed into that which did not exist, all
> things would have perished, since that into which they
> were dissolved would not exist.[4]

Epicurus's argument is psychologically related to what is called an *ad absurdum* proof in mathematics: the assertion one wants to establish is "proved" by showing that all the logical alternatives we can think of yield conclusions that are absurd. In seeking to establish the foundations of physics, Epicurus showed that on the assumption of a slow disappearance of matter, everything ultimately disappears (including the questioner!). At the other extreme, a genuine but slow increase in matter would ultimately produce a world that was *utterly packed*—and the questioner wouldn't be able to move a finger.

Over the centuries, various verbal garments were used for the not-to-be-questioned verbal anchors of science. A few of these were first principles, fundamental dogmas, self-evident truths, eternal truths, and what Thomas Aquinas called natural law. To the scientific mind, ecclesiastical anchors suffer from being too defiant, too uncompromising. The linguistic anchors employed by science are apologetic and somewhat hesitant.

Why? Because in science, occasionally a principle that has long seemed to be fundamental is found to be in error. As the 19th century turned into the 20th, it was found that mass could be changed into energy. Fortunately, in less than a decade, Albert Einstein showed how the two older conservation principles (of mass and energy) were combined by nature into one. From that time on, a new analytical entity, *mass-energy*, was known to obey the conservation law. Physics again had its anchor; compulsive scoffers had to look for other grounds on which to exercise their ambitions.

They did. After Einstein's new synthesis, dreamers of perpetual motion continued to multiply and bombard the U.S. Patent Office with proposals. None of them panned out, and finally the government office issued a ukase: no more applications for perpetual motion patents would even be looked at until—outside the Patent Office—the nonconservation of energy was established in the face of the most critical attacks.

Very narrow-minded of the Patent Office, you may say, but necessary for institutional sanity. Radioactivity shattered some of the scientific anchors but not the need for them. The institution called *science* has no authority like the Catholic Church's Vatican to appeal to. Fundamental statements are arrived at by informal consensus, and they are never asserted to be infallible. Nevertheless, a degree of stability among the fundamentals is desirable. They need a name that implies stability without completely closing the door to change.

An unofficial solution has been offered for this need: the word *default*.[5] Its understood meaning is this: "attacks on long-successful basic principles require the expenditure of effort (always in short supply). Therefore, in default of substantial and convincing evidence, *the burden of proof* will be placed on all statements that contradict the one that is here deemed the dominant one."

The default positions of a science that has long been in the charge of first-rate investigators are few in number, easily expressible, and widespread in application. Paradoxically, their strong acceptance by the inner circle approximates the confident acceptance of traditional dogma by theologians. A telling illustration is found in a criticism that Einstein made of an older physicist, Max Planck, whom he much admired. Einstein tells us that his 61-year-old colleague, during the solar eclipse of 1919, "stayed up all night to see if it would confirm the bending of light by the gravitational field [of the sun]. If he had really understood [the general theory of relativity], he would have gone to bed the way I did."[6] (Einstein was 40 years old at the time.)

Building on a succession of modest default positions, science progresses. The conservation of mass-energy is one of the most important default positions of the physical sciences. In laying the foundations for a sustainable population policy, we will seek comparable default positions on which population theory may be erected. One of these positions is the one Tertullian asserted—the essential conservation of available human resources. What other default positions can we add to the list?

5

The Stormy Marriage
of Economics
and Ecology

How many times in a week do you see the word *shortage* in a newspaper or magazine? By comparison, when (if ever) have you seen the word *longage* in print? Most likely, never. The word *shortage* is welcomed because it gives growth-oriented producers (both manufacturers and philanthropists) all the excuse they need to expand productive facilities and profits. *Longage*, however, implies that we need to curb growth. The ostrich within us doesn't want *that*.

The annals of academia are rife with the splitting of disciplines. Long ago *philosophy* (the word means "love of knowledge") split into natural science and the remainder that we now call philosophy. Then natural science began separating into many different sciences (a process that is still going on). Occasionally portions of two specialties melded together to produce a new discipline that was less restricted in its subject matter than either of its parent sources. This happened early in the 20th century when parts of chemistry and biology were fused into biochemistry. As the 20th century drew to a close, another union began to take place, between economics and ecology. Rachel Carson's *The Silent Spring* paved the way for the alliance.

The words *ecology* and *economy* are both derived from the Greek root *oikos*, meaning "home" or "household." Both specialties deal with the rational management of a household; ecology (say the ecologists) is the more inclusive discipline since it recognizes that both the inhabitants (people) and the furniture (the earth) are important in determining what happens in the household. In the hope of legitimizing the love affair, a union was publicly proclaimed, which soon gave birth to a journal of its own, *Ecological Economics*.

▲ ▼ ▲ ▼ ▲

Ecology, as a member in good standing of the natural sciences, inherited—from the 3rd century B.C.—the mantle of Epicurus. The classic statement (see chapter 4) of the conservation of mass served as a model for other conservation principles, which became default positions of the sciences generally. The guiding mantras of ecologists—"Everything is connected to everything else" and "We can never do merely one thing"—reveal a deeply conservative bias. By contrast, some economists still glory in the mantra "Ours is a limitless world" and the chant, "Win!—Win!" These wish-fulfillment battle cries have been accompanied by a treasure trove of astonishing statements by economists and popular writers on economics. Box 5-1 displays a bevy of these seductive ostriches on the sandy beaches of economics.

In 1987, the World Bank convened a meeting of economists and their critics just to explore the limits of economics. As reported by Constance Holden, a respected veteran journalist for *Science* weekly: "the economists at the meeting rejected the idea that resources could be finite. Said one: 'The notion that there are limits that can't be taken care of by capital has to be rejected.' Said another (to the ecologists): 'I think the burden of proof is on your side to show that there are limits and where they are.' They were suspicious of well-worn ecological terms such as 'carrying capacity' and 'sustainability.'"

The last demand is a shrewd one; the "burden of proof" issue is always difficult to deal with. But if the same reaction were let loose in the natural sciences, what would happen to the various conservation principles? In the sense meant by the quoted economist, how would one set about *proving* the conservation

Population Ostriches
on the Sands of Economics

1. Nature imposes particular scarcities, not an inescapable general scarcity.
   —*Harold J. Barnett and Chandler Morse*[1]

2. There is no danger from the exhaustion of physical resources.
   —*Peter T. Bauer*[2]

3. The United States must overcome the materialistic fallacy: the illusion that resources and capital are essentially things which can run out, rather than products of the human will and imagination which in freedom are inexhaustible.
   —*George Gilder*[3]

4. The action most urgently needed in the world economy is for the stronger economies to be willing to accept higher levels of living.
   —*Paul W. McCracken*[4]

5. The limits to growth may be adroitly sidestepped just as they begin to loom menacingly.
   —*Charles Perrings*[5]

6. We have in our hands now—actually in our libraries—the technology to feed, clothe and supply energy to an ever-growing population for the next 7 billion years.
   —*Julian Simon*[6]

Box 5-1

of matter? Of energy? The best we can do is to show the ridiculous or unbelievable consequences of their *not* being true. So, after examination, we call the conservation laws default positions and move ahead with our work.

There may well be no more conferences that are exploratory in the same way as the one held by the World Bank in 1987. The new hybrid discipline of ecological economics seems to be drawing its principal strength from the younger members of both the parental disciplines. Older economists who refuse to accord default status to the idea of limits are apparently being phased out by retirement.

Most of the professionals in both economics and the natural sciences are inclined to avoid public controversy, believing that the facts should speak for themselves. But journalists, ever on the lookout for controversies (which sell publications), are sucked into giving unlimited publicity to oddball ideas. Unfortunately, journalists, who as a group are badly educated, are poorly equipped to judge importance. Some writers make little effort to understand what the specialists are saying.

Professional scientists and economists should take on the task of correcting journalistic errors, whether of substance or emphasis; but the professionals find their work too much fun, and so they tend to neglect this civic duty. Moreover, the institutions that employ them would not be likely to reward them for forays into the education of the public at large when they have been hired to advance the frontiers of knowledge. The end result: public education remains in the hands of the poorly educated. Hence the long press life of some superstitions: the existence of extraterrestrial invaders, the danger of dental fluoridation, the innocuousness of nuclear power—and the belief that the world has no limits for human existence.

Fortunately, in 1995, a commission of 11 distinguished leaders in the two sciences, headed by the Nobel laureate in economics Kenneth Arrow, issued a two-page position paper that admirably shows the essential unity of economics and ecology. Its title was "Economic Growth, Carrying Capacity, and the Environment."[7] Thus did the commission notify the community that they thought it was time for the two disciplines to grow together.

▲ ▼ ▲ ▼ ▲

A pause is in order at this point. Julian Simon's off-the-cuff claim (box 5-1) that world population could grow without limit for the next 7 billion years needs to be run through a computer. Albert A. Bartlett,[8] a physicist at the University of Colorado, did just this, using only a moderately powered desk calculator. It flashed "Error," indicating that, multiplying steadily at 1 percent per year for 7 billion years, the population would soon surpass 9.99×10^{99}, the limit of his computer. When this fantastic error was pointed out to him, Simon said, "Oh, I meant 7 *million* years."

Well, everyone makes mistakes. Let's not be niggling in our criticism. Let us, Professor Bartlett said, assume that 7 million was what Simon had in mind. Assuming the present world population of almost 6 billion and the recent rate of population growth of 1 percent per year, how long would it take for the human population to equal all the atoms in the universe? The answer is shocking: *just 17,000 years.*

Depending on when we think the human species began, 17,000 years is only about one-third or one-sixth of the time our species has been on earth. That's far short of 7 million years; and as for 7 *billion* years—well, Simon really should have taken an elementary math course before he said anything about the consequences of human reproduction.

Julian Simon (1933–1998) thought he could pull the fat out of the fire by implying that *billion* was really a typographic error for *million*. But straightforward math shows how colossal was Simon's ignorance of the inescapable mathematics of biology. In lectures, Simon delighted in snowing his audience with statistics; but scientists soon recognized that he was not, in the strict sense, numerate in his thinking.

(We must not forget that the *relatively* tiny number 10 to the 99th power, at which Bartlett's computer gulped and quit, is *more* than the number of atoms in the entire universe, which is usually estimated at about 3×10^{85}.)

▲ ▼ ▲ ▼ ▲

From Epicurus on—for the past 22 centuries—mainstream science has been irrevocably committed to conservative thinking. This means that there is necessarily a rigor in scientific thinking: No conservation of the elements of thought—no understandable equations—no science.

Scientists have long been committed to the conservation of matter and energy. Only once in 2,000 years has this major scientific position been threatened with abandonment: when radioactivity was discovered. But Einstein's celebrated equation soon showed how a new conservative position could be established and defended. So scientists continue to put the burden of proof on anyone who asserts a nonconservative position.

This scientific attitude moved into the everyday world when, early in the 20th century, the U.S. Patent Office refused to consider any more proposals for perpetual motion machines.[9] Any economist who comes up with another win-win scheme thereby shows that he or she still believes in the fiscal equivalent of perpetual motion machines. The seeming appearance of economic wealth from nowhere is due to faulty accounting, such as not acknowledging, for instance, that the energy-rich oil drawn out of the ground was put there by the biological capture of the sun's energy, coupled with geochemical processes that converted biological products into oil.

Nevertheless, *officially* economics is a conservative discipline. Beginning students are introduced to the basic default position, "There's no such thing as a free lunch." Unfortunately, a few people who are accepted as spokesmen for the discipline of economics preach a contrary sermon to the public. A sample of these heretics grace box 5-1. Simon's technique of presenting a plethora of statistics (while ignoring conflicting data) has been called *obstructive empiricism*. As a virtuoso of the technique, he was much quoted by the press.

His pronouncements were welcomed by business executives who wanted to ignore the costly side effects of their activities. Biologist Hugh Iltis has emphasized the resemblance of Simon's work to that of the popes in Martin Luther's time. Like them, Simon was in the business of selling indulgences to those who were planning to "sin."

Mainline economists could point out the many errors in Simon's work; unfortunately, they seldom do so, perhaps because they dream of the financial support that may be forthcoming from the clientele to whom Simon appealed. The assumption of obstructive empiricists is one of perpetual growth, which is accepted as a form of optimism. (Optimism, of course, is more welcome than pessimism.) But, a critic might ask: what if the human body were programmed to grow forever in weight? Would that kind of perpetual growth be a cause for optimism?

During much of human history, optimism was arguably more realistic than its opposite. There was always land beyond the horizon to be occupied. By whom? By the young, who, as in so many species of animals, have "itchy feet"; by the losers in competition in the home territory; and by the unusually ambitious, who work hard for possessions and power. For most people, for at least a limited time, it pays to be optimistic. The pessimists, who may eventually prove to be right, are generally forgotten long before the results come in.

Operators of a conservative temper attempt to draw up a balance sheet to see how well they are doing. The effort pays off in business and in science, but when we come to accounting for the human adventure as a whole, we encounter a new phenomenon: the *environment* (or whatever equivalent term is used) is not a static element in our analysis. Applied science—engineering—continually expands the *human* environment. The invention of boats opened up new lands for settlement. Dams and canals permitted the storage and movement of water to more farming acres. Breeding better plants and animals increased the productivity of the land. And various ways of increasing human productivity followed from the capture and redirection of energy by windmills, dams, and water wheels, and by the combustion of (successively) wood, coal, oil, and gas. Nuclear reactors produce more electricity (though the true long-term cost of safely sequestering the waste products is still not known). In effect, the environment actually available to human beings on our finite earth has been greatly expanded. It is no wonder that intelligent people should be reluctant to give up the idea that there is no limit to the humanly available environment, even though the earth itself is finite.

▲ ▼ ▲ ▼ ▲

For many generations the expansionist view has been much strengthened by the value-laden noun *shortage*. Whenever an imbalance developed between supply and demand, it was immediately spoken of as a shortage, which promised higher prices for whatever product was involved. This meant that it was worthwhile for enterprisers to find new ways of producing (or releasing) the product in question. The prospect of profits was a powerful motivator for enlarging the effective environment of the human race.

The word *shortage* was coined in 1868 by speculators in grain. Until 1975 the word *longage* apparently did not exist.[10] And no wonder. A balance between demand and supply *might* have been sought by decreasing the demand—but who would pay for such a solution? Solutions of that sort fall into two classes: (1) reduce the number of people making the demand or (2) persuade individuals to settle for a smaller supply per person.

Largely at an unconscious level, two sorts of objections were raised to the first approach. If it is proposed (1) that population be reduced, someone soon indignantly asks: "*Who* is to be liquidated?" or (not quite so brutally) "Who is to be forbidden to reproduce?"

Taking the other tack (2): "If the community has to do with less, will the speaker please set the example and simplify *his* or *her* life first?"

It scarcely needs to be said that there is no obvious way for people to make a profit out of trying to correct a longage of people or their desires. But trying to cure a shortage offers all sorts of chances for middlemen to make money. Under the circumstances, the logically possible term *longage* long remained unsaid. It suffered the most complete taboo possible: it was not even mentioned as a forbidden term. In 1982, the expert polling organization run by Daniel Yankelovich found that 52 percent of the general public endorsed the view that "technology will find a way of solving the problem of shortages and natural resources." The relevance of *longage* was not investigated; the pollsters may well have been ignorant of the term.

Unfortunately, few people are acutely aware of the following basic fact of human politics: in dealing with problems of human need

We can't cure a shortage by increasing the supply.

All we do is encourage the production of more people or greater demands. The shortage continues undiminished or is even increased. (Adding two more lanes to a highway, for instance, ultimately increases traffic jams.)

▲ ▼ ▲ ▼ ▲

We now understand the reasons for the stormy courtship of ecology and economics. The italicized sentence above, though firmly based on experience, strikes at what has, until now, been the very heart of the discipline of economics. (This heart is seldom or never explicitly mentioned by economists.) Economics offers correctives to shortages. Longages are someone else's business (and the less said about them, the better!).

Then there's the word *environment.* We begin by looking for this word in the index of important economics texts. We look first in Adam Smith's *The Wealth of Nations* (1776). *Not there.* Then John Stuart Mill's *Principles of Political Economy* (1848). Again, *not there.* Next, as a tiny sample of modern textbooks, we look at the very successful *Economics* by Paul A. Samuelson, first published in 1955 and revised through many editions. Both *longage* and *environment* are missing—at least through 1964. (Checking later editions is an exercise left to the reader.)

There is, however, a significant entry in *The Fortune Encyclopedia of Economics,* published in 1993, under the general editorship of David R. Henderson. *Environment* is not there either, but *environmentalism* is, with several entries. The treatment is adequate, except that it is pretty well segregated from the general body of economic thought.

What is the significance of the suffix *-ism?* The introduction of an *-ism* term is generally made by an unsympathetic observer. Religious "fundamentalism" was not given its name by a believing fundamentalist; similarly for the perpetrators of the terms *pacifism, militarism, populism,* and *nativism.* Crying "-ism!" these days is like crying "leper!" in biblical times.

Ultimately, the *-ism* may be detoxified, so to speak, and the purified term may find use in the general literature. Actually, by 1993 quite a few books written by economists presented a favorable view of environmental matters, but encyclopedias are generally behind the times. The production of such a book is necessarily a joint effort. Two heads may be better than one, but in the acceptance of intellectual revolutions, 10 heads are slower.

▲ ▼ ▲ ▼ ▲

Dipping into the great bulk of economic analyses, you soon discover that *environment* is usually no more than a ghost in the woodwork. Using italics to indicate its ghostly character, we can write it in as {Env} in a generalized ecological-economics equation thus:

{Env} + Resource + Process ⟶ Product + By-products + *{Env}*

Since the by-products may be salable, it is desirable to list them. If they are not salable, "we throw them away," economic apologists say. Following this policy for centuries showed that, in the real world, {Env} was anything but a ghost. Mine tailings have made deserts out of hundreds of thousands of acres; sewage effluent can kill the fish for miles downstream; thrown-away trash may spoil many square miles of wetlands on the shore, making them unfit for the breeding of valuable fish; the devastation created by the clear-cutting of forests can make impossible the regrowth of the complex beauty of a true forest while satisfactory profits are being extracted from biologically impoverished and esthetically repulsive "tree farms."

Taxpayers should be (but are not) upset by the fact that bulldozed access roads in an abandoned forest (paid for mistakenly in the name of private enterprise by an overfriendly government) cause the loss of millions of tons of topsoil. Equally disturbing are the radioactive waste products left over from the generation of electricity from thorium. These *by-products*—what a dismissive term!—will, since we haven't yet figured out what to do with them, threaten humanity for *thousands of years*. Yes, thousands of years. Such wastes can be watched over by a conscientious government, but what government—since the time of the pharaohs—has endured for even a few centuries? In the absence of caretaker stability, radioactive wastes threaten unknown and anonymous multitudes of our posterity.

The environment, whether named in equations or not, has been central to ecological thinking from the very beginning— long before Rachel Carson brought the ecological revolution to public notice. By contrast, the literature of economics included hardly a mention of the environment during the first two centuries after Adam Smith.

To understand the revolution now in the making, two generalizations should be kept in mind:

1. What passes for true Smithian economics assumes that human behavior can be understood in terms of completely egocentric motivation.
2. By contrast, ecological economics assumes egocentric motives tempered by a concern for posterity.

▲ ▼ ▲ ▼ ▲

We see only what we have names for.

Sadly, this brutal statement is too true. The word *shortage* has long permitted humanity to accept uncritically the idea of nearly infinite resources always available to be dipped into. As humanity stops complaining of shortages, we can anticipate that other mantras, beloved of ecologists, will be welcomed into economic discussions. These should include:

1. "When you try to pick out any thing by itself, you find it hitched to everything else in the universe." Source: John Muir (1911), a pioneeer ecologist.
2. "There's no away to throw to."
3. "We can never do merely one thing."

These statements all say pretty much the same thing. Every proposal to reform the details of our world is sure to set in train processes that many people are not emotionally prepared to deal with. Every proposal to build a dam, to widen a highway, to cut down another forest, to turn wetlands into salable real estate, or to bury unwanted waste products is sure to have *unintended consequences*, to use the pungent phrase introduced by sociologist Robert K. Merton in 1936.[11] The innocence of the childhood of humanity is at an end. From now on, we must accept responsibility for *all* the unintended consequences while doing our best to predict them in advance; and avoid them— generally by truly conservative action, which has not been fashionable for the past two centuries.

The public should be warned of the ambiguity of the word *conservative*. Political conservatives and commercial conservatives repeatedly recommend social action that is not, by any stretch of the term, *environmentally conservative*. In such cases, the recommenders stand to profit personally from the destructive action. Forests are razed and harmful dams are built to conserve and extend the fortunes of rich men and the reputations of politicians.

Understandably, many leading economists have not welcomed the introduction of the total environment into the developing portrait of the human condition. Since Adam Smith's time, economists have pretty well established themselves as the designated spokesmen of a greatly simplified human household. After 1962, the controversial atmosphere was embittered by scores of books and hundreds of articles taking one side or the other in the argument. A few economists urged their colleagues to accept the new broadening of the field. For instance, E. F. Schumacher's *Small Is Beautiful* (1973) put the argument in terms his colleagues could understand, whether or not they agreed with him. Twenty-two years later another economist, Wilfred Beckerman, hurled a fresh missile at the biologists. In a sarcastic nod toward Schumacher (now dead), Beckerman called his book *Small Is Stupid.*

Often bystanders felt that the adversaries on both sides were wrong. Perhaps Schumacher's book should have been entitled *Too Large Is Ugly*, while Beckerman's response should have been called *Too Small Is Stupid.* Both books were unbalanced, but, of course, rhetorical extremism sells books better than temperate language. Extremism appears to lead to clear-cut decisions, whereas moderation embarrasses us by emphasizing problems that are yet to be solved.

The deceptively infinite character of the human environment rapidly shrinks when our attention shifts from the resources to the "sinks" that are used for the disposal of wastes. The more successful a continuously growing population is in extracting wealth from nature-the-resource, the sooner it will suffer from the intransigence of the unintended creation, nature-the-sink. It begins to look as though *sink* considerations will move more minds than *source* considerations.

6

Consequentialism
Nature's Morality

The purpose of this brief chapter is to distinguish between two controversial ideas that are all too often confused in the public mind: evolution and natural selection. The chapter that follows develops at greater length the properties and importance of the second idea.

The idea of biological evolution was a natural outgrowth of the idea of progress, a gift from ancient times. Philosophers tended to see a program in progress, a sense in which what was happening was directed toward a future that was somehow known to some invisible spirit. In the early days of its existence, evolution also was thought to be the acting out of a preestablished program. This assumption was abandoned when it was realized that only *ex post facto* could one "see" any program in biological evolution. People find it easy to make sense of what *has* happened, while they wisely hesitate to predict what *will* happen.

The future of our population—the future of humanity—is under a cloud of evolutionary uncertainty. Yet to do nothing is not a realistic option because *nothing never happens*. The uncertainty is both a threat and a promise. Which of the two will dominate our future is—theoretically—within our power to determine. Surely we are the only species of animal that is aware of this opportunity. Can we—or some of us—correctly discern the dangers of the future? And can a minority of our population persuade the majority to grasp the nettle of responsibility for what will happen?

Repeated polls show that something like half of all Americans doubt the truth of biological evolution. Professional biologists are amazed at this large proportion: it is as though most people thought Newton's gravity was pure superstition. In the case of the biological theory, however, people probably reject it for what they think are its hidden (and repulsive) implications.

Strictly speaking, the idea of evolution is primarily a historical idea, referring to the past. The entangled concept of natural selection is a scientific idea, and it refers primarily to the future. Whenever we raise a question about choices to be made in the future, we are raising an ethical question. It is always tempting to evade a discussion of choices by implying that an assertion about the past necessarily dictates a particular choice of future action.

Recently, for instance, a woman named Mrs. Seagraves, testifying against teaching evolution in the California schools, said: "If you teach that man is an animal, then there's no right or wrong. What would you expect him to do except start breaking the laws?"[1] Earlier, the poet W. H. Auden had introduced a variant of this argument: "All attempts to account for our behavior on the basis of our pre-human ancestors are myths, and usually invented to justify base behavior."[2] Notice that neither of these critics addresses the question "What is *true*?" Both of them think that evolutionary truth would produce wicked behavior. Their conclusions can rightfully be said to be driven by *motivational ethics*. Those who, enamored of vituperative rhetoric, are trapped in motivational thinking are almost solely concerned with their interpretation of the past. (It feels so *good* to castigate your opponent for a past that neither you nor he or she can undo!)

But scientists, anticipating the future, favor *consequentialist ethics*, which is less interested in historical origins and more concerned with the future consequences of present acts. By focusing on the alterable future and forgoing the pleasure of calling their opponents names, consequentialist ethicists may, in time and with careful reasoning, be able to lead to agreements on policy. Conclusions derived from the younger consequential ethics are often incompatible with those dictated by the more ancient and rigid rules of motivational ethics. (Consequential ethicists are, of course, often accused of being amoral.)

To judge by press reports, the long-standing conflict over evolution is primarily a dispute about the past. Is it true that all living things were created in a couple of days (as reported in Genesis)? Or did they evolve over millions of years? "Listening with the third ear" of the psychoanalyst, one hears many complaints like the ones cited above by Auden and Mrs. Seagraves. Since the future can be altered (and the past cannot), our need for a good consequentialist ethics exceeds by far our need for a historically accurate motivational ethics. In its major thrust, Darwin's *On the Origin of Species* is a consequentialist treatise on the power of natural selection. Darwin showed that natural selection is a consequence that follows automatically from the consistent fecundity of variable species reproducing in a competitive world of limited capacity. (These fruitful assumptions came from Malthus—who failed to deduce natural selection from them.)

For resource constraints to produce differential and lasting effects on the survival of variants within a species, there must be a mechanism for originating new types of individuals and a mechanism that makes the differences, at least in part, inheritable. People have long been guided by the folk saying "Like produces like." At the same time, it was recognized that it is perfectly natural for a small percentage of newborn animals to be freaks, that is, different from the parental strains. Unfortunately, in Darwin's day, the folkish view of heredity was unable to reconcile these two ancient discoveries.

As it happened, the elements of the genetic theory Darwin needed were published by Gregor Mendel in 1866, but in so hesitant a manner that neither Darwin nor any other biologist of standing knew and understood what Mendel had written. Not until 1900 (41 years after publication of *On the Origin of Species*, 18 years after Darwin's death) did three biologists reactivate Mendel's theory. It rapidly became a cornerstone of the biological sciences.

▲ ▼ ▲ ▼ ▲

The solidity of a scientific theory is shown in its ability to reconcile data that appear (at first) to lead to irreconcilable conclusions. This point is nicely illustrated by two examples of animal behavior that seem at first glance to imply utterly contradictory ethical principles but that, in fact, spring from a unified set of principles, those of consequentialist ethics.

First case: the European swift.[3] Like other swallows, the European swift gets all its food on the wing, engulfing insects flying in the warm air. When it is trying to raise a family, the swift needs extra food. If the weather turns cool, insects disappear from the air and the bird cannot satisfy its natural demands. Suppose, at this point, the bird finds that its life is encumbered with a nest of eggs that need hatching. Later, of course, the hatched youngsters will need to be fed by the parents. What if it seems that the overly cool weather is going to persist straight through to this "later" time?

In this case, the economical thing to do is to get rid of the eggs and try again—later—when the weather improves. Why *economical*? Because if the weather doesn't improve, the mother bird will waste part of the all-too-brief summer in sitting on eggs the fate of which is problematic. If she always acts to conserve that which has already been produced, on the average she will produce a reduced number of clutches during the breeding season. She can have more offspring *in the long run* if she "liquidates" eggs that are doomed by the coming cool weather.

In any case, as the cool weather continues, the swift tips one egg out of the nest; then a second; then a third. With *zero* fertile eggs—with zero commitment to an unknown future—she is now free to start a new gamble on the later prospects of her germinal cells. Like a human weather prophet, the bird may be either right or wrong. All she can do is play the odds, which inherited behavior equips her to do.

If it makes a motivational moralist happy to say so, he or she can report that the mother bird sometimes *murders* her potential children, hoping that she will thereby increase the total number of progeny she leaves behind when she departs this vale of tears. (At this point an irate reader may exclaim: "I see where you are going with this argument, and I won't have it! It is never licit for a human mother to kill her offspring. We aren't birds! We don't lay eggs that require external care! We don't eat mosquitos! The bird's problem has no relevance to human problems." Perhaps, perhaps not. But, before committing ourselves, let's enlarge the ethical universe with another case drawn from nature.)

Second case: a cricket species.[4] Richard Alexander has shown that one particular species of cricket behaves in the following way. The mother cricket lays a large number of eggs. When the young hatch, one might expect many of them to die young. (Great fertility in our resource-limited world is generally accompanied by great mortality during the earliest stages.) This mother cricket, however, has a trick up her sleeve, so to speak. She offers herself as the first meal of her babies, thus increasing the probability of their survival. They eat her alive. Literally. No doubt a mother cricket who makes this one-time sacrifice leaves behind more descendants than would one who turned her children loose to shift for themselves. At any rate (if the reader will pardon the loose language), "natural selection thinks so." (Similar behavior occurs in several species of spiders.)

Reconciliation. Many a human moralist would say that the mother bird sometimes commits *murder*. The same commentator would likely praise the mother cricket for the nobility of her *suicidal* self-sacrifice. The contrasting behaviors can easily be reconciled intellectually. Natural selection is the feature that unifies the apparently contradictory behaviors.

Natural selection invests in success; it cannot do otherwise. Focusing on the future and ignoring the past, consequentialist ethics tells us that the two cited behaviors have the same consequence: the production of more posterity in the long run.

In contrast, actions that human beings are pleased to call *charitable* often prove to be *investments in failure*. The news commentator Eric Sevareid was right when he said that "Most problems are caused by solutions"—*would-be solutions*, we should say.

▲ ▼ ▲ ▼ ▲

With these instances of nonhuman behavior, we have arrived at a fork in the road of ethical analysis: some readers will choose one path, some the other. Perhaps we need to consult our friendly Martian, who is (by hypothesis) wholly rational and not bound by the commitments of Earthlings. Unmoved by socially inherited emotions, our Martian evaluates the behavior of all earthly animals—swifts, crickets, human beings, and all the rest. The moral taken in the discussion of avian ethics is often called that of *relativistic ethics*. Reading terrestrial magazines and newspapers, the Martian would soon discover that this term is now a denigrative one on earth. "Interesting!" he would say.

Pursuing his studies further, he would find that the term *situational ethics* is also used, thus calling attention to the importance of the existing situation in determining the conclusion reached. This term evokes the same sort of criticism as *relativistic ethics*.

Then there is the term *consequentialist ethics*, which is favored here because it easily evokes the important idea of natural selection. But though more than a century has passed since Darwin convinced most biologists of this fact, the idea is still disturbing to many other people. However, it needs to be wrestled with.

More important than the term selected is the attitude of the speaker to evocative words generally. Here there can be a sharp split between scientists and the general public. I have previously characterized the attitude of scientists toward words as follows:

> A scientist cannot accept the orientation of the first sentence of the book of John: "In the beginning was the Word, and the Word was with God, and the Word was God." No doubt this statement can be interpreted in terms of symbols, parables or myths, but all such substitutes for real propositions are ambiguous. Scientists are more attracted to the motto of the Royal Society of London: *Nullius in verba*. If I were charged with altering Scripture to conform with science I would say: "In the beginning was the *World*, which everywhere and forever envelops us; against this external reality all human words must be measured."[5]

For nearly two millennia, the attitude presented in the book of John has been accepted as the socially correct one. But what sort of man was responsible for this verbal presentation? In essence, he was a priest. For thousands of years the priesthood greatly influenced people's perception of reality. The prestige of priests has now passed in large measure to professional word-smiths—journalists, essayists, prominent public speakers. The exposure of almost all of these people to science is minimal. They all demand too much prestige for their words. Many of the paradoxes perceived by them disappear when semicompetent words are replaced by the mother ideas resident in a wordless World. If the ignorance of science among influential word-smiths becomes much more widespread, we shudder to think of the future of civilization.

▲ ▼ ▲ ▼ ▲

What kind of a world would it be if its inhabitants lived only by a motivational ethics that, looking to the past, frequently demanded an investment in failure? Does not the great wealth of literature on human ethics need to be reexamined in the light of Darwinian insights? Can the human species long survive if it habitually invests in failure?

It surely has not escaped the reader's attention that the behavior of mother crickets is not *totally* different from the behavior of human mothers and fathers. Some human parents, mothers particularly, sacrifice much of their well-being to give their children a better start in life. If you don't believe that state-ment, try justifying your position to middle-income parents who have accepted the burden of financing a good (and expensive) college education for their children.

Following an old tradition of textbooks, I present some ques-tions for the reader to answer:

1. If the crickets and birds presented here had the gift of human speech, how would each species justify its behavior?
2. What would be the attitude of highly intellectual crickets and birds toward "Tertullian's blessing" (chapter 2)?
3. Reexamine the arguments for and against contraception and abortion in the light of consequentialist ethics. Are there other human actions that may be said to constitute "investments in failure"?
4. Can a community make a better future for its descendants by invariably subsidizing failures?

7

Natural Selection
God's Choice

In the development of population theory, the issues of evolution and selection—both natural and social—come up repeatedly. To make sure that we are approaching a question with the right tools, we must first be clear about the difference between scientific and historical inquiries. A model scientific question is this: "Is there a gravitational attraction between two massive bodies in space?" If at any time in the future someone doubts our answer, he or she can test it anew. Again and again, if necessary.

Now consider these questions:

Did the Orientals invent the stirrup?
Do people and chimpanzees have common ancestors?

In contrast to strictly scientific questions, these two are essentially historical. Their answers lie in the past, which cannot be recovered. Naturally, we have more confidence in scientific answers than we do in historical ones.

Failure to distinguish these two classes of statements has produced unnecessary conflicts in America during the 20th century. Strictly speaking, the second italicized question should, like the first, be dealt with in the history departments of universities.

It is ludicrous that religious fundamentalists have fought to have their version of evolution taught in the public schools under the name of *creation science*. Science it is not; as *creation history* it might have a place in public school history classes (though historians might object). In following the arguments of the present work, the reader should keep in mind the following:

Natural selection is a scientific question.
Evolution is a historical question.

However, because of the technical competence required—carbon dating, for instance—the study and presentation of evolution is routinely assigned to biology departments. Tradition wins out over logic.

The confusion can be traced back to the publication of Charles Darwin's *On the Origin of Species*.[1] The book could just as well have been entitled *The Evolution of Species*. In both cases, the title suggests (in the strict sense) a historical subject. Darwin did not include the word *evolution* in his original treatise. It was not until the 5th edition (10 years later) that he inserted—awkwardly—a few references to evolution.

The fact is, the roots of the idea of biological evolution go way back. The ancient Greeks occasionally toyed with it. The French scientist Henri Lamarck wrote a book about it published in the year of Charles Darwin's birth. And P. T. Barnum, that shrewd judge of human curiosity, displayed "missing links" in the side shows of his circus 17 years before publication of *On the Origin of Species*. Long before the 19th century, as Arthur O. Lovejoy has shown in *The Great Chain of Being* (1936), scholars habitually placed the various species in a chain. Formally, this was just a logical chain, but many amateurs assumed that the chain indicated family relationships, sometimes referring to its "missing links."

The subtitle of Darwin's book announces the appearance of a scientific thesis: *By Means of Natural Selection, or the Preservation of Favoured Races in the Struggle for Life*. The author's consuming interest was in the mechanism he was proposing as a major cause of evolution. In fact, for several years before he sent his book to the publisher, he proposed that its title be simply *Natural Selection*. He was talked out of this by friends, who felt that the public would find such a title puzzling. Perhaps they were right.

▲ ▼ ▲ ▼ ▲

Most people have an approximate idea of how we should go about testing the validity of scientific statements. Does gravity exist? Turn off the fuel to an airplane in flight and note what happens. However worded, a scientific statement refers to facts that can be tested in the future. But because historical conclusions refer to a past that is beyond recall, they can easily lead to enduring disputes. As a matter of political reality, *people who dislike particular historical conclusions cannot be decisively silenced.* Let three examples of this unwelcome generalization stand for many.

First, in recent years, emotional assaults have been launched against the generally accepted belief that there was indeed a Nazi Holocaust before and during the Second World War. A minority of cynics completely discount the vivid testimonies of hundreds of people who swore they had escaped from brutal prison camps.

Second, most of the people who speak European languages see a historical unity and progression of their kind of civilization beginning in Egypt and the Tigris-Euphrates basin and spreading slowly over Europe before being further extended by colonization into the Americas. In recent years, a small, racially led group has asserted that the major ingredient of Western civilization actually came from black people living south of Egypt. This position has been criticized by the columnist George F. Wills in the following words: "They believe that the truth of a proposition about history is less important than the proposition's therapeutic effect on the self-esteem of people whose ethnic pride might be enhanced by it." If so, history becomes a verbal soporific. (In this case, should not the therapeutic effect of history be evaluated by the Food and Drug Administration?)

Third, two decades before Darwin published his world-disturbing book, Nicholas Wiseman, cardinal of London, dismissed the hypothesis of the evolutionary origin of man by contending that "It is revolting to think that our noble nature should be nothing more than the perfecting of the ape's maliciousness."[2] An older friend of Darwin's, Adam Sedgwick, later used his prestige to muster the troops to oppose both the assertion of the antiquity of man and his "ignoble" origin, but he decided to grant Darwin half a victory:

> I can no longer maintain the position which I have hitherto held. I must freely admit that man is of a far higher antiquity than that which I have hitherto assigned to him. But, Gentlemen, I shall always protest against that degrading hypothesis which attributes to man an origin derived from the lower animals.[3]

Sedgwick's attitude is common among many who still reject the theory that humans are descended from less noble animals. One cannot but feel compassion for the heartache they experience as the biological view becomes more common. It is painful for an adult to restructure the framework of his or her beliefs. For society as a whole, the time required for this process is often measured in generations.

If the generalized scientist has a religion, it is summarized in the injunction "Never suffer a delusion to live!" So how should we react when the following three propositions are passionately recommended to us?

1. There was no Nazi Holocaust;
2. The European source of Western civilization is a fraud;
3. The human species did not evolve from nonhuman animals.

Since I have been trained in science, I have nothing *authoritative* to say about the truth of any of these historical propositions. Insofar as I understand how professional historians do their work, I presume that the propositions are false. But I will listen to other opinions. In the meantime, as a member of my community and a concerned grandparent, I will object strenuously if publicly supported schools teach any of the three propositions.

An author is keenly aware that every clear-cut statement he or she makes will alienate some potential readers. But if the author says nothing, clearly he or she risks having no audience at all. Unconventional views are best introduced slowly, with adequate evidence. I am willing to see my potential audience contracted to the group that agrees with me on the above three propositions. Now I want to move on to other propositions that are still properly debatable. It has, I think, become increasingly obvious during the past generation that the idea of selection and its consequences, as well as other insights growing out of economics and ecology, have a real bearing on basic principles of ethics. *Some* of the new conclusions are, when first faced, shocking to *some* people. The controversial interactions of ecology, economics, and ethics need to be explored, along with the truly penetrating implications of the findings.

▲ ▼ ▲ ▼ ▲

How would you answer a poll taker who asked you, "Do you believe in God?" In a recent survey, 26 percent of the general public answered "Yes" to the question; 12 percent answered "No"; and the remainder scattered themselves throughout a forest of ambiguous "Maybe's."

The fact is, of course, that the unitary word *God* covers a large population of gods of differing abilities. There is God the Originator of All, God the Intervener in Daily Affairs, the God Who Answers Prayers, the God of Justice, and so on. When all the polling is done, we have scarcely a clue as to what the report's summary, "26 percent believers," means. So the finding is not very useful.

Most of the world's many gods are providers, parental figures who are presumed to play much the same role as that of human parents during the years of childhood. The idea of God's providence, or divine providence, is an old one, but the word *providence* did not receive much publicity until the Enlightenment, when many people became embarrassed to use the word *God*. *Providence* then became a fashionable substitute. Naturally, with the passage of time, the rationale for using the substitute became obscure, as is obvious in a passage from *The Nine Tailors*, a mystery story by Dorothy L. Sayers:

> "We mustn't question the ways of Providence," said the Rector.

"Providence?" said the old woman. "Don't yew talk to me about Providence. I've had enough o' Providence. First he took my husband, and then he took my 'taters, but there's One above as'll teach him to mend his manners, if he don't look out."

The Rector was too much distressed to challenge this remarkable piece of theology.[4]

It might be objected that this passage is, after all, only a matter of fiction and hence undeserving of serious mention. But good fiction writers often take small pieces of dialogue from real life. (The fiction comes in when such passages are stitched together.) In any case, the theology of Sayers's old lady is no more remarkable than the statement of a real minister of the cloth, testifying in Arkansas at an antievolution trial in 1981: "It is possible to believe that God exists without necessarily believing in God."[5] (What box should a pollster check for *that* reply?)

The meaning of the Enlightenment of the 18th century is found in its elevation of reason above faith. On the way to formulating a definition of God that reason could approve of, the philosopher Gottfried Wilhelm Leibniz (1646–1716), meditating on the wonders of the physical world, asserted that "God himself could not choose without having a reason for his choice."[6] Enlightened investigators were, then, engaged in discovering God's reasons for assembling the amazing world that surrounds us.

Two centuries later, Albert Einstein put the idea on a more personal basis. As his biographer Banesh Hoffman has explained: "When judging a scientific theory, his own or another's, he asked himself whether he would have made the universe in that way had he been God."[7] During most of the Christian era such a procedure would be deemed sacrilegious, but not, I think, by many scientists. In justifying the inescapable default positions of science (see chapter 4), it was made clear that they serve as a way of bringing the infinite regresses of logic to a close. The word *God* can be seen as an early attempt to verbalize the idea of a default position.

It is easy to understand why many people of deep faith were greatly disturbed by the developments of the Age of Enlightenment. Yet heartache was not a necessary consequence of the change in people's mental furniture. Shortly after *On the Origin of Species* was published, the minister Charles Kingsley wrote to thank Darwin for his copy of the book. In his letter, Kingsley broke rank with most of his fellow ministers by giving evolution preference over the biblical story of Creation, saying: "I have gradually learned to see that it is just as noble a conception of Deity, to believe that He created primal forms capable of self-development into all forms needful . . . as to believe that He required a fresh act of intervention to supply the *lacunae* which He himself had made. I question whether the former be not the loftier thought."[8]

The Enlightenment can be viewed as an unconscious drive to remove a quasi-personal Mover from the explanation of the world. In the 18th century, many persuasive writers pointed to the evident "design" in nature. One of these writers was William Paley, who, at the turn of the century, said that where there was a Design, there must be a Designer. In the 19th century, the philosopher Henry Sidgwick chose to see law as the organizing principle of the world, saying (in effect) that where there is Law there must be a Law Giver. In such a climate of opinion, it was natural that Darwin's emphasis on natural selection should lead his detractors to assert that his work merely created a new causative formula: where there is Selection there must be a Selector. Thus one sort of Darwinian critic hoped to transform a threatened intellectual revolution into a minor palace revolt. (Darwin did not accept the suggestion.) Another sort of critic, in an essay published in the *Princeton Review*, asserted that Darwin's *Origin of Species* "avowedly and purposely ungods the universe."

The Reverend Kingsley saw great good coming out of this ungodding. This minister (and writer of novels) was gifted in making deep ideas understandable to children: his *Water Babies*, published four years after Darwin's *Origin of Species*, reported that a child came to Mother Nature, expecting to find her very busy, and instead found her with her hands folded. She said: "I'm not going to trouble myself making things. . . . I sit here and make them make themselves." Thus, in figurative language intended for children, Kingsley hinted that God may not have given us a ready-made world, as described in chapter 1 of Genesis; rather, his primary gift was the *process* of natural selection, which, given enough time, could—and did, and still does—put together the incredibly complex and beautiful world we live in.

Ninety years later, Watson and Crick showed how nature starts the trick with DNA (or one of its chemical predecessors).

▲ ▼ ▲ ▼ ▲

And how do things manage to make themselves? The basic key is natural selection—the differential reproduction of competing variants, resulting in the production of more offspring by the variants that are best fitted to the environmental challenges and fewer offspring produced by less-suitable variants. *Fitness* has no absolute measure; it is always *fitness for a defined environment.* Change the environment, and fitness must be redefined. With many environments, many hereditary types become possible. To achieve a diversity of types, there must be a matching diversity of environments to act as preservers of the hereditary differences produced by mutation and rearrangement of genetic elements. For a diversity of types to survive, individuals must be protected from simplistic selective factors.

The number of different species now living in the world is not precisely known, but the leading estimates lie between 10 and 50 million. Kingsley's apparently passive Mother Nature has accomplished a great deal by just keeping her hands folded and letting natural forces, including natural selection, do the work. Of course, there's been plenty of time: several billion years, it appears.

In establishing the conservation of matter as a default position, Epicurus explained how the only alternatives we can imagine lead to ridiculous conclusions (shown to be ridiculous by their absence in fact). Natural selection is in an equally secure default position. New variants are always being produced by the ordinary processes of chemical change and biological reproduction; in that variegated landscape of biological types, variants that are better fitted to any particular environment necessarily—*as a matter of definition*—reproduce themselves more abundantly in that environment. (One cannot imagine the truly less fit reproducing in larger numbers, because they would then, by definition, be more fit.)

Natural selection, in Darwin's sense, necessarily works only with inheritable differences. If the differences are not inheritable, selection (of a more ordinary sort) may still be effective; but if the difference is not inherited via the genes, its preservation cannot be ascribed to natural selection.

Notice that natural selection is only one element of a large class of selections. The more general principle is this: *We get whatever we reward for.* If that wording seems too personal, we can say: *The reward determines the outcome.*[9]

If the laws of society reward for bank robbing, society will get more bank robbing. If our methods of winnowing candidates for high positions favor stupidity, we will get stupid politicians. If we reward for athletic prowess, we will get great athletes. If we reward the lazy, their tribe will increase. (Whether heredity is involved in any of the differences involved is of only secondary importance: more important in the long run than in the short run. But sometimes there is no long run.)

The rest of this book is dedicated to showing how fantastically effective selection is, whether it is natural selection or merely social selection. Glimmerings of an understanding of selection are found in the literature of 2,000 years ago, but the earliest insights were often effaced by ostrich like behavior. Now that we have a good understanding of the role of genetic elements in producing individual differences, we face a greater difficulty: will we find ways of bringing together (synthesizing) individual differences into social amalgams that can survive the inherent disharmonies that turn up in mixtures? The natural sciences have probably made it possible for millions—*probably not billions*—of human beings to live sustainably on the earth. Will the still embryonic social sciences find ways to turn potential survival into actual survival?

More to the point is this fact: the selective value of selfish behavior is easy to appreciate. What is more difficult to understand is how individual behavior that serves the needs of *others* in a population can be favored by selection. To this topic we now turn.

8

Altruism

Natural selection is an inescapable default position of all biology, and as such calls for no experimental proof. Would human beings with six fingers per hand be superior to those with five? If this were so, six-fingeredness would soon be the norm of the species. In truth, we deduce natural selection from whatever exists. By definition, *inferior* (as determined by the environment) cannot replace *superior*, whether we like it or not. With our domesticated animals and plants, *we* define *superior* in a way that suits us and then select for a constellation of genes that might not long survive under human-free, wild conditions. (High-producing milk cows haven't a chance of survival when turned loose in a wild area with predators, diseases, and competing species of ungulates.)

(Very few predecessors of Darwin fully appreciated the similarity between natural selection and the selection practiced by animal breeders. Therefore, they did not explore the consequences of natural selection. Darwin had many precursors who developed the idea of evolution, but no one really explored the parallel concept. Genius is a mysterious thing.)

Purposely closing their eyes to the default nature of natural selection, many people become quite emotional in condemning a "dog-eat-dog world," "nature red in tooth and claw," and so on. Such emotional reactions oversimplify the real world. Every species is confronted with not simply *one* environment but many. A stupid human being who is unusually athletic can prosper; so also, in a different way, can a bright but clumsy person. The philosopher Herbert Spencer used the phrase *survival of the fittest* to epitomize the result of the selection process. Two historical aspects of this phrase deserve mention: (1) Spencer publicized it seven years before publication of *Origin of Species*, and (2) Darwin never liked it, merely mentioning it in later editions of his magnum opus.

There's something aggressive and intolerant about the concept of *survival of the fittest*. It poorly mirrors the richness of the world we live in. *Fitness* is an immensely complicated concept, involving as it does the simultaneous measure of two extremely diverse sets of variables, the first determined by the nature of the individual, the second by the demands of the environment. For a long time to come, social theory will be reacting to the puzzles created by the interaction of these two variables; in the meantime, every proposed solution should be regarded with suspicion. Nevertheless, natural selection must be a major default position of our theorizing.

▲ ▼ ▲ ▼ ▲

About 50 years before Christ was born, the Jewish sage Hillel said: "If I am not for myself, who will be for me? And if I am for myself alone, what am I? And if not now, when?" Long before Hillel, people must have been aware of the conflict between actions narrowly focused on the self and actions contributing to the good of others. But not much systematic progress could be made until names were available for the opposing impulses. The best of the terms—*egoism* and *altruism*—appeared in print only a short while ago. The earliest use of the English word *egoism* (from the Latin *ego*, meaning "I") recorded in the *Oxford English Dictionary* dates from 1722—about 140 years before *natural selection* appeared in print. The earliest use of *altruism* (from the Latin *alter*, meaning "other") did not appear until 1853, and its birthplace was France.

That *egoism* should have been coined before *altruism* should not surprise us: the competition of one person with another inevitably impresses on the party of the first part the fact that *other people* think of themselves first. (Secondarily, with some reluctance, Ego finally surmises: "Perhaps I too am like that.")

In a world of mixed types, clearly (we think) the wholly egoistic type will have an advantage over egoism-plus-altruism types because the latter always run the risk of weakening their influence by strengthening the positions of more egoistic competitors.

This view was supported by a wealth of statements by theorists and practitioners of what is thought of as Smithian economics. In 1854 Hermann Heinrich Gossens put a religious spin on the egoistic impulse: "Going against self-interest only inhibits God's plan. . . . How can a creature be so arrogant as to want to frustrate totally or partially the purpose of his Creator?"[1] In 1881 F. Y. Edgeworth dispensed with religion and said simply that "The first principle of economics is that every agent is actuated only by self-interest." (The *only* is in the original.[2]) In our own day, Bill Gates, thought to be the richest man in America, is reported to have said that he "distrusts anyone who is only moderately greedy." He is an authentic intellectual heir of Adam Smith.

It is easy for a biologist sitting in an armchair to "prove too much" as he or she evaluates natural selection operating on behavioral genes. In the hard-nosed view a simple argument may overwhelm us, convincing us against our will that altruism is impossible. What does the hard-nosed argument leave out?

The most obvious omission is parenthood—both the fact and its consequences. Recalling the story of Alexander's cricket (see chapter 6), ask yourself if natural selection could have selected for suicidal sacrificial behavior *if*, in each case, the mother cricket succeeded in saving the life of *only one* baby cricket in her whole lifetime? Geneticists say the clear answer is "No," because only half of the baby's genes come from the mother (the other half coming from the father); so a hypothesized gene for altruism would have only a 50 percent chance of surviving. Selection would work *for* a mother cricket who saved her own life rather than that of a *single* baby. Two babies are the "breakeven point." When three or more infant lives are saved by the self-sacrifice of the mother, natural selection will work in favor of the sacrificial gene.

Altruism—active concern for others of the same species—is an obvious characteristic of social species like *Homo sapiens.* Our remarkable success in competing with other forms of life is derived from brain power. The explanation of this evolution includes several elements.

First, the brains of the "higher" mammals are very complicated and delicate, the human brain most of all. The function of the skull is to protect this delicate computing instrument. Rigidity protects. The schedule of human development is "written" (in the genes, of course) so as to push the development of the brain along faster than that of the rest of the body. As a result, babies and young children have outsized brains—the head is about one-fourth the length of the whole body rather than the one-seventh that characterizes the adult head-to-body proportion. (Some of the early Renaissance paintings look strange because the baby Jesus has a head that is only one-seventh of the body length. This is the adult ratio.)

Second, to get a good start on brain development, the accelerated schedule begins before the baby is born. This creates problems for the mother; the uterus and the birth canal have to be highly distensible. A conflict in selection is created: the baby's survival is favored by a large head, while the mother's survival is favored by a small-headed baby. As usual, opposing selections produce a compromise.

Third, after birth, the life and development of the child are made possible by the altruism of the parents. This is one of the many examples of *kin altruism,* the force of which varies directly (but approximately) with the closeness of the kinship. The extraordinary social power of kin altruism is obvious in the domestic honeybee.

It has long been common knowledge that the working caste of honeybees protects the hive by stinging interlopers, though the act of stinging sometimes removes the stinger bee's viscera, causing its death. At first glance, this looks like sacrifice of the purest sort. Natural selection should work against the persistence of stinging behavior, but it doesn't. Selection works *in favor of suicide.* How can that be?

Worker bees are reproductively sterile; whatever genes they

may have can be passed on only by their fertile brothers and sisters that possess the same genes (even though they may not be expressed). So the suicidal genes are transmitted to future generations by individuals who never commit suicide.

This may seem odd until we note that human beings are protected from microbial disease by phagocytic cells in the blood that attack microbial invaders even though the phagocytes themselves may be killed by the invading cells. In any case, the phagocytes of one generation never lead *directly* to phagocytes in the next. Reproductive cells—eggs and sperm—are the carriers of self-sacrificing phagocytosis, though they themselves are never so "foolish" as to behave that way.

Among bees, the entire colony can be viewed as constituting a single organism—or, as it is sometimes called, a *superorganism*. The hereditary behavioral genes of sterile workers are subjected to natural selection via the reproduction of fertile brothers and sisters that possess the same genes, though their behavior does not reveal their genotype.

The phagocytic cells in our body are exact analogs of the worker bees in the "body" of the superorganismic bee colony. In a deep sense, all genes act selfishly in their own interest, even when (as in these two cases) they seem to be acting altruistically in the interests of the parent "body" of which they are a part.

This situation was understood by Darwin and was greatly elucidated in the century following the publication of the *Origin of Species*. Several biologists were involved in this work, one of whom, Richard Dawkins, explained it well in his book *The Selfish Gene* (1976). Altruism, which otherwise might seem a bad investment of the individual, is "financed" by the ties of kinship.

Before going further, note the following implications for human behavior. *If* membership in a family were always determined by a random assignment of foster children to adults, *then* there would be no way that altruistic genes could be selected for. No Dawkins would arise to sing the praise of selfish genes. Without the effective protection of individuals by truly biological parents during a long period of maturation, the human species would have to settle for a greatly truncated period of development. The fact that genes are necessarily selfish creates a payoff for sacrifices less "expensive" than suicide if the sacrifice benefits the close relatives of the one making the sacrifice. The closer the genetic relationship, the better the selective payoff.

(Only if we were to become the domesticated animals of some other species—say, the Superwomen of Saturn—could our young be taken care of with different final effects. With no superspecies in sight, our species has to incorporate a measure of kin altruism in its makeup. Conceivably this association might be confined to the period of childhood; but apparently it is easier for selection to work in a crude way to favor altruistic behavior that extends to the adult period of both parents and children. There are, however, great variations among the different human cultures—a fact we should not lose sight of.)

Nongenetic factors can play a role, too. The policy "You scratch my back and I'll scratch yours" can create exchanges between individuals not known to be related by family: between friends, for instance. Unequal exchanges may create a "debt" that militates against the habit, but when times are prosperous, such debts are more readily tolerated, with little loss to the more giving member of the pair. As a generalization, rich communities are somewhat less selfish than poor ones, though there are exceptions.

Be that as it may, there is little doubt that folk ethics is inclined to pronounce altruism and its consequences good, while egoism and its effects are seen as at least mildly reprehensible. All this could be predicted by our Martian friend, who is completely objective in his observations of Earthlings. The Compleat Egoist is well advised to openly and loudly praise altruism (in others) while failing to call attention to the egoistic element in himself. Call this inconsistency hypocritical if you wish; it certainly produces a well-greased path to social power, be it in the state or the church.

Once the social position of altruism is well established, the impulse toward altruism can easily serve egoistic purposes. We should remember the remarks of the helpful devil in *The Screwtape Letters* by C. S. Lewis:

> A sensible human once said, "If people knew how much ill-feeling Unselfishness occasions, it would not be so

often recommended from the pulpit"; and again, "She's the sort of woman who lives for others—you can always tell the others by their hunted expression."[3]

It is easy for Ego to acknowledge the ubiquity of egoism—*in the other person*. Realistically, we must admit that considerable psychological strength is imparted to individuals by their unquestioned assumption of a large amount of altruism in themselves.

As evidence of how passionately individuals may cherish a belief in their own altruism, consider the implications of this press dispatch from Omaha, Nebraska, on 26 October 1978. A would-be volunteer blood-donor—a determined altruist, be it noted—after being rejected for technical reasons by the center's medical director, went on a rampage, killing two persons (including the director) and wounding several others. Such behavior is hardly what a folkish understanding of ethics would predict. But the following conclusion must be accepted as fact: *the refusal of a gift can evoke aggressive reactions in the would-be giver.*

Once we recognize that Lewis's devil was right in recognizing the paradoxical truth that altruism could be a weapon of egoism, we conclude that the classic behavior of the Kwakiutl Indians of the American Northwest was, after all, not very different from that of the rest of the world. Their potlatch was a ceremony at which a rich Indian gave much or all of his wealth to his tribal friends—even, sometimes, to strangers. Many Europeans, when they first learned of this behavior, believed that the Kwakiutls were so weird that they must belong to another species. It was hard for *Homo sapiens europiensis* to admit that any race could behave so unsapiently and yet still be considered *H. sapiens*. Western economic theory holds that competition must be principally concerned with acquiring wealth, not with giving it away. So passionate is the belief that money competition reigns supreme that some conventionally educated people fail to realize how strong can be the desire to be well thought of by one's associates.

Once we focus on the word *competition*, we suspect that it is at least conceivable that a distribution system could relinquish the European type of competition in favor of a system devoted to the generous dispersal of wealth. After all, the Bible tells us that

"It is more blessed to give than to receive." Why then do we not take this injunction seriously and build our economic system on it? The Europeanized mind boggles at the suggestion. Furthermore, pessimistic critics can point out that some "high-minded" motives can be just as greedy as simple greed for money. C. S. Lewis would probably agree.

Most (all?) religions praise altruism unreservedly. At least equally encouraging is the fact that in the last two decades, a large number of scientific theoreticians have moved into the egoism-altruism field of investigation to see if they can't resolve its internal inconsistencies. They have made significant strides in a short time (though much remains to be done). Only some of the high points of this work are ready for simplification.

▲ ▼ ▲ ▼ ▲

At the outset, it is clear that altruistic actions must be subject to scale effects. I may cheerfully share my food with my neighbor, but I am unlikely to be much concerned with the even greater needs of a person 12,000 miles away. I *discriminate* between people on various grounds: kinship, companionship in work and play, and so on. Furthermore, distance is important in at least two ways: in reducing the reliability of the reports of need, and in eroding the quality and quantity of the material aid that is sent halfway around the world. No one classification of altruisms is official, but that shown in figure 8-1 will serve to make the necessary points. Pure egoism is included in the list for completeness; it so seldom occurs that it may be of no importance in social theory.

Scale effects fall into two categories. First, there are those effects that decrease with the size of the category: these include trust and the feeling of affection. The shaded area on the right side of figure 8-1 gives a plausible estimate of the amount of "loyalty power" inhering in each level of the graded sequence. But a scale effect of the opposite sign produces the "political power" gradient shown on the left side of the figure. The larger the membership of a given category, the greater the quantity of power that it can exert on the individual, through police action, and so on.

Since these two kinds of power vary differently in response to the size of the group, we cannot assume that a political organization that works well in a nation of one size will work equally well in nations of all sizes. Now comes a vital and strongly tabooed question: can a nation that has functioned well under one political system recognize when growth has gone on so long that it is time to change its political organization? No persuasive answer to this question has yet been worked out, though the growth of nations in the past century has been without precedent in history. And the population continues to grow in almost every nation.

Traditional ethics is the ethics of a village. In such a limited arena, one is pretty sure how much of a neighbor's misfortune has been earned by incompetence and how much is due to the undeserved slings and arrows of outrageous fortune. A good person, identifying psychologically with his or her neighbor, is apt to say: "There but for the grace of God, go I" while helping the neighbor with some friendly work. (In a true village, alms in the form of money are seldom encountered.)

The exact shape of the shaded areas in figure 8-1 is not known. Nevertheless, experience tells us that there is some sort of inverse relationship between loyalty power and political power. The source of loyalty power is essentially biological: in the history of the individual, it is derived from the realities of kinship. We feel that loyalty to one's relatives and longtime

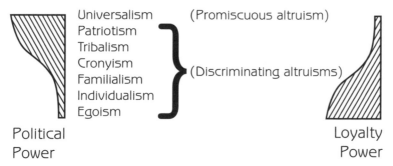

Figure 8-1. The hierarchy of egoism and the various altruisms. The hatched areas indicate the approximate power of each kind. Taken from G. Hardin, *Living Within Limits* (New York: Oxford University Press, 1993).

associates is essentially noble. But when there is an attempt to extend this sort of loyalty to much larger units of association—businesses and nations—conflict develops between the loyalty of intimates and the loyalty to large, impersonal groups. The stage is set for nepotism, favoritism, and lawbreaking.

To keep large groups honest, there must be whistle-blowers who inform the general public that intimate associates are violating publicly agreed-upon rules. But whistle-blowers are often ruthlessly punished socially by their close associates—coworkers, club members, families, and so on. Anticipating this possibility, most potential whistle-blowers never raise the warning whistles to their mouths.

Looking over the roster of some 180 nations, an objective observer must reluctantly conclude that the majority of the world's governments are corrupt. Egoism can be socially destructive; so can selective altruism. Political science desperately needs its Newtons and Darwins to set civilization firmly on a better path to survival.

9

Coercion

If you want to understand the nature and consequences of coercion, don't expect much help from a dictionary. A typical definition of the verb *coerce* is as follows: "*tr.v.* To force to act or think in a given manner; to compel by pressure or threat." The emphasis on force satisfies the lawyers, who apparently use the word in no other way. For emotional people, lay or lawyerish, the word *coercion* has, as the political scientist William Ophuls admits, "a nasty fascist ring to it." But Ophuls is not satisfied with this condemnation. Some qualification is needed, because (he says) "political coercion in some form is inevitable," and the act "is an inextricable part of politics, and the problem is how best to tame it and bend it to the common interest."[1] Yet probably not one person in a hundred has realized that coercion is inescapable.

Many are the attempts made to soften the impact of the dictionary meaning of the word. Not uncommonly, *voluntary* is married to it: the result is an oxymoron. It's the sort of reality the 20th-century gangster Al Capone dealt with when he said: "You can get much farther with a kind word and a gun than you can with just a kind word." Mario Puzo, a novelist dealing with the gangster world, repeatedly has characters in *The Godfather* say: "I will make you an offer you can't refuse." Perhaps the illusion of voluntary compliance is needed "to keep the peace" (of a sort).

If we suppose that it is only criminals who crave the support of this illusion, what are we to say of the insight of William Paley, whose works were so important in Darwin's education? In 1802 Paley wrote: "Money is the sweetener of human toil; the substitute for coercion; the reconciler of labour with liberty."[2] Paley's courage failed him: money is not a substitute for coercion; it *is* coercion, albeit of an acceptable kind. Money is both the kind word and the gun. If the firepower isn't great enough to persuade the targeted person, increase the offer. But at whatever level a bargain is struck, our economy is built mostly on the "voluntary coercion" that takes place in the exchange of money for other things. Once we understand the role of money, *coercion* loses its nasty ring. The subject of economics may, in fact, be defined as the discipline that deals with acceptable coercions.

If the word *coercion* meant no more than "compulsion" or "force," it would hardly be needed in the arsenal of our rhetoric. But the coercion here described might better be called *reciprocal coercion* or, by mutual agreement, simply *coercion* in a nonforceful sense. It is the sort of behavior the French political scientist Helvétius (1715–1771) must have had in mind when he said:

> The whole art of the legislator consists of forcing men, by
> the sentiment of self-love, to be always just to one another.[3]

The result is government by the majority. This is probably the least onerous coercion we can hope for, but it is not always successful. The 18th Amendment to the U.S. Constitution (prohibiting consumption of alcohol) was passed by the majority of the legislators in a representative democracy, but a mere majority proved to be not enough for this controversial measure. In just over a decade, the amendment was set aside by another amendment. The numerical ratio between majority size and acceptability is not a simple one. Intuitively, we suspect that the more emotional the issue, the greater is the power of a minority to hold reformers at bay.

▲ ▼ ▲ ▼ ▲

The remainder of this book will discuss possible reforms in the organization of society. Almost inevitably the average reader— *whether he or she is in favor of or opposed to the change in question*— will first assume that the text proposes that force be used to achieve the reform. On the contrary: the two issues—goal and path—need to be kept separate. It may be that no force-free path leads to the goal, but this should not be gratuitously assumed beforehand.

The effect of scale is, of course, an important thread that runs through the tapestry of change. A change that is possible with a small number of participating citizens may be difficult or impossible with a larger number. Experiences of the Amish people bear this out.

A young Amish woman explained to a visitor why her church frowns on central heating systems. Said she: "A space heater in the kitchen keeps the family together. Heating all the rooms would lead to everyone going off to their own rooms."[4] Thus do these admirable people maximize their *other-directedness,* to use David Riesman's term. As one of the Amish put it: "Everybody knows everything about everyone else." The speaker did not complain. (Many non-Amish would.)

Anecdotes can be helpful in visualizing political problems. If negotiations are not carefully conducted, violence may follow. A fascinating account of such an outcome during the reign of Charles III of Spain has been retold by Will and Ariel Durant:

> [The Marchese de' Squillaci] stirred up a revolution by attempting to change the dress of the people. He persuaded the King that the long cape, which hid the figure, and the broad hat with turned-down rim, which hid much of the face, made it easier to conceal weapons and harder for the police to recognize criminals. A succession of royal decrees forbade the cape and the hat, and officers were equipped with shears to cut the offending garments down to legal size. This was more government than the proud Madrilenos could stand. On Palm Sunday, March 23, 1766, they rose in revolt, captured ammunition stores, emptied the prisons, overwhelmed

soldiers and police, attacked Squillaci's home, stoned Grimaldi, killed the Walloon guards of the royal palace, and paraded with the heads of these hated foreigners held aloft on pikes and crowned with broad-brimmed hats. For two days the mob slaughtered and pillaged. Charles yielded, repealed the decrees, and sent Squillaci, safely escorted, back to Italy. Meanwhile he had discovered the talents of the Conde de Aranda, and appointed him president of the Council of Castile. Aranda made the long cape and wide sombrero the official costume of the hangman; the new connotation made the old garb unfashionable; most Madrilenos adopted French dress.[5]

Question: Who was being more coercive—Squillaci or Aranda? More to the point: is it not more productive to abandon the *polar* classification ("coercive" or "not coercive"), replacing it with a *graded* classification (like the catalog of colors in the rainbow)? With this in mind we can replace the *go/no go* dichotomy of today's dictionaries with the following default position:

All persuasion takes place through coercion.

The persuading act may be as gentle as a sweet young thing's "Pretty please!" or as savage as an official's lash in Singapore, but the object of attention is offered a choice. (Al Capone may not have been far wrong after all.)

▲ ▼ ▲ ▼ ▲

A century before the illuminating episode in Spain, John Locke distinguished three levels of coercion. Some men, he said, becoming aware that they have broken God's law, may mend their ways when they imagine the divine punishment yet to come. Others, believing they have broken a human law, may anticipate more worldly punishment. But, says Locke, we find that the greatest part of mankind "govern themselves chiefly, if not solely, by [the] law of fashion; and so they do that which keeps them in reputation with their company, [with] little regard [for] the laws of God, or the magistrate." Locke's final conclusion bears emphasis: *"No man escapes the punishment of the company he keeps."*[6] If an offender cannot endure the punishment for an act he clings to, he must change his company.

Fear of disapproval is the major force that keeps a society intact: fear of God, fear of the police, and fear of the judgment of neighbors. Religious authorities want the fear of God to be the predominant controller. Civil authorities want fear of police and courts to dominate. But, says Locke, the opinion of one's neighbors trumps all others.

Intuitively, it should be obvious that the ability to escape society's punishment is directly related to the density of the community's population. The isolated heterodox behavioral act is harder to detect in a large community, as God and the bench lose power. Is this why so many people apparently prefer to live in cities? On the other hand, those whose beliefs are heterodox can more easily find, and preferentially associate with, others of their own persuasion in a large city. The result is a behavioral fragmentation of society, which unquestionably affects the feeling of loyalty toward the whole. Those who think that Locke has placed the three major coercive forces in the correct hierarchy cannot but wonder whether a civilization can long survive with an official scale that assigns so large a role to fashion.

▲ ▼ ▲ ▼ ▲

The story from the court of Charles III suggests a solution to a crucial problem of our time. (At least the problem is seen as crucial by those who meditate on the systemic disorders that have surfaced violently in the multicultural Russias, in multicultural central Africa, and in the multicultural Balkans during the last two decades.) One of the measures needed to put a stop to America's race toward multiculturalism is a system of reliably identifying those inhabitants who claim the protection of laws designed for a unicultural world. Some people think that we need a reliable identity card.

Paradoxically, Americans have already made a good beginning in devising such a system in the Social Security card and multiple credit cards, but taboo dictates that we avert our eyes (and our minds, and our language) from the possibilities of what we already have in hand. An identity card is opposed by a large, mixed company: by recent minority entrants who don't want to risk being counted; by some (not all) religious groups;

by some (not all) liberals; by some (not all) conservatives; and by the politically powerful American Civil Liberties Union. Though several European nations have used—with minimal abuse—identity cards for several generations, American folk wisdom says it can't be done here. Ambivalent church doctrines don't help, and legal traditions are confusing.

But there is always fashion, which, used imaginatively, can move practice to new levels. It should yet be possible for a charismatic leader—with the help of a growing minority of followers—by a vigorously publicized example, to make the carrying of a self-evident identity card fashionable here.

▲ ▼ ▲ ▼ ▲

When Ego thinks a change needs to be made, it is up to her to try to persuade others. After a while a movement develops, and it inevitably evokes a countermovement. If the struggle equilibrates at a level near one of the extremes, a law may be passed. Perfection is not possible—not in this world, at any rate. In a world far more populous than that of the Amish, Hélvetius was probably right: it *is* probably solely through the enactment and enforcement of human laws that we can come close enough to realizing our ideals.

In a stabilized nation there may still be a minority that dislikes a particular regime. Those who dislike what is agreed to by the majority have to consent to being coerced into obeying the law. To insist on unanimous agreement would be to make all forms of government impossible. The formula for a working society of our sort is quite simple:

Mutual coercion mutually agreed upon.

Do you still bristle at the word *coercion*? If so, you evidently do not recognize the above expression as the formula for a representative democracy or republic. To condemn the coercion of the individual by the group is to reject democracy.

10

Diseconomies of Scale

Ostrich Myopia

In the investigation of altruism (chapter 8) we found that the size of an interacting group affects the behavior of its members. In social animals generally—certainly among *Homo sapiens*—size must always be taken into account in predicting a group's behavior. The ancient Greeks were well aware of this: in recommending democracy as the preferred form of government, they were thinking of about 5,000 active citizens. (Their theory arbitrarily ignored slaves and partially disenfranchised classes.)

In political theory, size must figure into any default position: size of population, size of political units, magnitude of social powers—whatever. Economics has readily grasped the importance of *economies of scale*—economies to be gained by producing *more* of some product. For instance, expanding automobile production from 10,000 cars per year to 1 million markedly reduces the cost per manufactured unit because the capital cost of expensive machinery is divided among more of the units produced.

But what are we to say if we find that the costs—however reckoned—per unit of production or per unit of service *increase* with size? The "infrastructure" of a transportation system runs into this problem. Double the number of cars on the road and you increase the time it takes people to get to work (and "time is money"). But real estate developers, who have great influence with the press, do not encourage the mention of the "downside" of what they are doing.

In transportation matters, society repeatedly runs into *diseconomies of scale*. Perhaps we hope to escape them by expanding four lanes to six? But because of cloverleaf interchanges, and so on, roads with greater capacity cost more per unit built than do roads of lesser capacity. Moreover, the loss of land to the servicing of transportation drives up the price of real estate. Result: nearby homes cost more. In addition, settling for less land around each house creates needs that must be met in other ways; for example, home gardening must be replaced by more trips to the store.

Urbanization has other disadvantages that are seldom dealt with productively. The frequency of many kinds of crimes is greater in a large city than in a village. Whatever the sociological reasons for this diseconomy of scale, the relationship is a fact.

How easy it is for ostriches in our state legislatures to close their eyes to reality is shown by an experience in the state of Utah. The legislature passed a law requiring property taxes to be reduced as the value of property rose with crowding. The inescapable result: a steady increase in the debt of local governments. Then other sorts of taxes had to be increased.

Realists who, abandoning the ostrich stance, rudely mention diseconomies of scale are called pessimists. The 1933 edition of the *Oxford English Dictionary* does not contain the word *diseconomy*. However, the word is included in the 1972 Supplement, with 1937 given as the earliest known year of use. Significantly, the American *Encyclopedia of Economics*, though it was published as late as 1993, has *no* index entries for "scale, diseconomies of," though it has five under "scale, economies of." With no sign of a cessation of population growth, diseconomies of scale are sure to become a "growth industry" (of sorts). Our economic ostrich suffers the myopia of head-in-the-sand disease. In its biased mind, to grow larger is normal; to grow smaller or stay the same size is pathological. "Grow or die!" repeat the real estate developers of Ostrichland.

Biologists are astounded at this defect of economics. Only a minority of true organisms seem to grow indefinitely, for example, some whales, which live in a supporting medium that does not impose a scale effect on their major functions. But with most land-dwelling animals, the numerous metabolic "programs" are subject to different scale effects; the final size is programmed into the zygote. Development involves a period of maturation that continues until the individual reaches a genetically programmed maximum; *then it stops*. (Acromegalic giants of circus sideshows evoke our pity; we would be revolted at human growth that continued until a height of, say, 30 feet was reached.) The more complicated the system—organism or society—the more essential is it that it be programmed to respond to the command "So far—and no farther!"

The myth of the normality of perpetual growth is popular among business promoters and economists. Yet, in many instances, people who cling to this belief may help turn myth into reality—*for a time*. This is the grand illusion that brings riches to real estate developers and other chamber of commerce types. This is the myth with which Henry George broke his pick with his *Progress and Poverty* in 1879.[1]

Tragically, *in the short run*, economic rewards favor those who believe in the illusion of perpetual growth. Worshippers of growth, by planning their actions in accord with the illusion, prosper from what Robert K. Merton in 1948 called the *self-fulfilling prophecy*.[2] (The social sciences are different from the physical sciences.)

The error of fatalism is often assumed to be but an error of the past. But the quotation in box 5-1 by George Gilder shows that an ancient fatalism can be as seductive as ever. In a best-selling book, Gilder asserted that resources can have no limits because they are "products of the human will and imagination which in freedom are inexhaustible." Everybody loves freedom, so that's that.

▲ ▼ ▲ ▼ ▲

A question that Gilder raised (though he apparently did not know it) is this: with respect to a possible optimum and in the absence of a thorough analysis, what is the proper assumption to make? Conventional economists imply: "The maximum is best." This position is strengthened by the ease with which maximum points can be calculated mathematically. The mathematics is easy if you are dealing with only one independent variable at a time. But in the real world, many variables are involved in any important operation; maximizing one of them affects (often adversely) unnumbered others. Maximize the speed of automobiles and you increase many other things that you don't want to increase: the rate of accidents, the stress of driving, the cost of buying a home, and the crowding of people, with its concomitant loss of intimate personal contacts as the craving for privacy increases. Partial differential equations deal with such problems, but the result is seldom satisfactory to everyone involved. The assumption that passes the ecological test for a default position is this:

The maximum is not the optimum.

The quality of life (however defined) is affected by many variables. With many alternative "solutions" to partial differential equations, how do you settle for a single one? *Quality of life* opens up a Pandora's box of complexly interrelated problems. But life will never be satisfactory so long as Madam Ostrich, put in charge of the human household, is allowed to sweep problems of scale under a sandy rug.

▲ ▼ ▲ ▼ ▲

Fatalism, implicitly claimed but not openly acknowledged, sells many a controversial idea. In 1993 Jorge G. Casteñada, a professor of political science at the National University of Mexico, told an American audience that mass immigration of Mexicans to the United States was "inevitable, necessary and highly desirable."[3] The second and third assertions imply hypotheses that can be tested; the first assertion constitutes a fatalism that is untestable. If fatalism is a matter of faith, the fatalistic position is likely to win, in part because of a lack of frankness in revealing the motives behind the assertion. This is evident in the account of an experience of the American physicist Albert A. Bartlett:

Many years ago I was discussing the population growth of Boulder with a prominent member of the Colorado legislature. At one point he said, "Al, we could not stop Boulder's growth if we wanted to!" I responded, "I agree; therefore, let's put a tax on the growth so that, at a minimum, it pays for itself, instead of having to be paid for by the existing taxpayers." His response was quick and emphatic: "You can't do that; you'd slow down our growth!"[4]

Sadly, most people who might have an experience like Bartlett's would not detect any inconsistency in this justification of growth. Nongrowth is viewed as pathological, unnatural. But when we really look at nature, we discover that the situation is quite otherwise.

To take only one example among millions, consider the height of the human body. At birth it is about 50 centimeters; at one year of age, about 70 centimeters—an increase of 40 percent. If the human body continued to increase at a rate of 40 percent per year, by age 20 the person would be 1,394 feet tall. (Surely it isn't necessary to belabor the point by calculating the human height at 60 years of age?)

Nor should it be necessary to point out that the perpetual growth of individual human bodies would create daunting societal problems in the construction of suitable houses, autos, and aircraft to fit the resultant "diversity." (This last remark should remind the reader how thoughtlessly diversity is praised these days.)

In nature, not only is it normal for the growth of the whole to be subject to limitations, it is also normal for the growth of parts to come to an end. In a course in human embryology, one of the first things the premed student is surprised to learn is that in the development of the human embryo, three different kidneys are produced. Kidney 1 appears, grows for a while, and is then liquidated. So also with kidney 2, which is succeeded by kidney 3 (with which we are born). No doubt there are profound biochemical reasons for this "wasteful" development, though we don't know what they are. But the point is undeniable: *de*development can be as normal as development.

Again: a baby is born with a fairly large thymus gland in the chest region, but this gland later shrinks; by adulthood it has usually disappeared. This example is extreme, but every tissue and organ has a period of rapid growth, followed by a near-cessation of growth in the adult—except cancerous tissues, of course. (But who wants cancers? Doctrinaire economists, perhaps?)

The various functional units of a political organism—a nation, for instance—can be considered as so many "tissues" that are more or less integrated into a unitary body. When a political organism grows, it tends to keep the tissues that have served it well in the past. Its political leaders seldom consider that scale effects may be operating.

Understandable mechanisms account for this sort of conservatism. For one thing, the brains of the young furnish them with no other memory. For another, what Locke called the *law of fashion* (chapter 9) impinges on other-directed citizens. Thus, the New England town meetings that functioned so well in a village when only a few hundred people were entitled to vote were continued for quite a while after the village was no longer competent to deal with its burgeoning numbers. Time-honored verbal formulas were accepted as gospel.

Eventually, of course, the simple democracy of small numbers was replaced by the representative democracy we call a republic. And now? We've gone beyond this point in our nation, though the rhetoric is still not far from that of a simple democracy. The citizen with a request went first from an encounter with equals to a face-to-face dialogue with his or her legal representative. With further growth the citizen had to confront 50 or so legislative aides, hoping that one of these busy young persons would find time to carry his or her message to the representative.

Looks familiar, doesn't it? (Kidney 1⟶kidney 2⟶kidney 3⟶*what next?*) Whatever the arrangement, the result will be called *democracy*. (If you doubt that, carefully observe the rhetoric of developing countries that are only now emerging from totalitarian governments. Democracy, whatever else it is, is today's fashion. Fashion in speech interferes with the perception of reality.)

▲ ▼ ▲ ▼ ▲

Animal populations are governed by limitations. When rabbits were introduced into Australia, their populations at first grew explosively—but not forever. Economically threatened human beings imported diseases to help control the rabbits. Mutations and selection among both rabbits and disease organisms eventually produced a new equilibrium (of sorts).

Every growth phenomenon that exhibits economies of scale in its early stages must eventually run into barriers of diseconomies of scale; these either bring growth to a halt or extinguish the structure entirely. From its earliest days, ecology has acknowledged the existence of both economies and diseconomies of scale, of generation and extinction. But economics, because of its commercial usefulness to greedy humanity, built some major theories on only half of the truth. Professional economists are now in the process of reforming their discipline; they are not helped in this necessary labor by the insistence of profit-seeking laymen that they focus their best thinking *only* on economies of scale.

We have yet to hear of a politician who was elected to office by reminding constituents of diseconomies of scale. Nature, however, will not forever let the voters forget.

▲ ▼ ▲ ▼ ▲

Perceiving an economy of scale often requires only the simplest observations. Not so with diseconomies, because the behavior of two or more forces is involved. An example from a human organization can show why an ecological view is essential.

A couple of decades ago, the managers of the Volvo factory in Sweden became concerned with the high cost of production of their cars. Fractionating and simplifying the motions of individual workers had already produced the maximum efficiency of production (so long as workers were content to be little more than living extensions of unfeeling machines), but unpredictable worker interruptions—strikes and slowdowns—had led to diseconomies of scale.

The managers became convinced that, by reversing course and permitting—even encouraging—workers to shift from one job to another, the price of the cars could be reduced. Their hunch was right.

Conscientious students of strikes are convinced that the demand for higher wages is often a cryptic call for work that is more challenging or more emotionally interesting (choose your own adjective). The impersonality of a production line can alienate workers. Alienation can produce revolt. And revolt, when opposed, often focuses on the wrong cause. Wages may be raised beyond necessity. Each inefficiency of production has the potential to breed others. A vicious cycle is set up by society's blindness to the psychological environment of work.

Labor relations should be labeled as problems in *human ecology*, a discipline in its infancy. Making the total society more efficient requires insights far beyond those demanded of accountants and economists, *sensu stricto*. Replacing classical economics with ecological economics opens exciting vistas of the future of humanity.

▲ ▼ ▲ ▼ ▲

One more projectile remains to be aimed against the advocates of growthmanship. Among this throng, the great god Growth may well command more worshippers than the Judeo-Christian God. When at last a considerable number of critics made the Growth worshippers doubt the salability of their faith, the first reforms were largely semantic—trying to use the opponents' rhetoric in the defense of the god Growth, which was going to solve all of society's problems. A flurry of papers and books praised the beauties of "sustainable growth," ignoring this critical fact:

"Sustainable growth" is an oxymoron.

From its Greek roots, *oxymoron* means "pointedly foolish" or perhaps "wise-foolish" (e.g., "mournful optimist"). In a world that is, for all practical purposes, finite, *sustainable growth* is truly an oxymoron. (Of course the perpetrator of an oxymoron hopes that his or her auditors do not perceive the internal contradiction.)

On the other hand, *sustainable development* can be defended because (for instance) an adult can continue to develop his or her intelligence for many decades without any growth in body weight. But in the economic world, *growth* and *development* have for so long been near-synonyms that we should always be suspicious of the praise of sustainable development.

As far as nature is concerned, every species fits into an environmental niche that, in effect, dictates the best blueprint of its body. By natural selection, a species comes closer and closer to developing the ideal blueprint (which does *not* include "Grow forever" as a commandment). Of course, the development of one blueprint-dictated species can change the environment significantly for other species, which then "seek" better blueprints. This endless process, called *coevolution,* is only now being researched sufficiently. But it would be unwise to use the ambiguity of any current position as an excuse for worshipping *growth forever.*

In summary: it's easy to make money by generating more supply; it's very difficult to make money by persuading people to be more economical in their demands. For almost two centuries, half of the meaning of the word *economical* has ignored the subjectivity of the word *demand.* The omission disturbed ecologists; it delighted most economists and businessmen. Accepting the fact that the world available to human beings is a limited one will be one of the most difficult tasks ever tackled by our species. The intermediate costs will be high; the ultimate reward will be survival itself.

11

The Dream
of One World

No discussion of the future of the human population is complete without a careful description and evaluation of the dream of One World. The hope that humanity might someday be united into a single world sovereignty was broached in Judeo literature as early as the 5th century B.C. Castigating unholy priests, Malachi (chapter 2, verse 10) asked: "Have we not all one father? Hath not one God created us?" Malachi undoubtedly meant "*all* of us"; he hoped to extend the family feeling at least to an entire village, the members of which are apt to be relatives anyway. Some Greek philosophers soon included the entire human species in a universal kinship. But can we safely extend this figure of speech to the 5 billion people now alive? Or even to the hundreds of millions in the largest nations? A scale effect is involved in the answers to these questions.

Rhetorical "brotherhood" is immensely appealing. After all, family loyalty is the most intense loyalty that most people ever know. Down through the ages we find example after example of serious thinkers who founded their systems of ethics on rhetorical kinship. In the 3rd century B.C., Zeno of Cytium (one of the founders of Stoicism) dreamed of a day when all the warring states "would be replaced by one vast society in which there would be no nations, no classes, no rich or poor, no masters or slaves; in which philosophers would rule without oppression, and all men would be brothers as the children of one God."

Moving swiftly to the Christian era, we find Erasmus in the 16th century proudly announcing, "I wish to be called a citizen of the world." Two centuries later, the eloquent journalist Thomas Paine (who was nominally loyal to both Britain and revolutionary America) announced: "My country is the world; my countrymen are mankind."

In the 20th century we find that the most influential propagandists for One World are, like Paine, journalists and literary people. In 1901 H. G. Wells said flatly that a world state was inevitable. In 1920 he closed his best-selling *Outline of History* with these words: "Our true God now is the God of all men. Nationalism as a God must follow the tribal gods into limbo. Our true nationality is mankind."

During the Second World War, the magazine editor Norman Cousins said that man "shall have to recognize the flat truth that the greatest obsolescence of all in the Atomic Age is national sovereignty." Dorothy Thompson, an influential newspaper reporter, said flatly that "There must be a world state." In 1945, Mortimer J. Adler, whose opinions placed him at odds with many of his fellow philosophers at the University of Chicago, went further: "We must do everything we can to abolish the United States."[1] We must replace patriotism, Adler said, with love of a single world sovereignty. The same conclusion, but with a different emphasis, was reached by the environmentalist Lester R. Brown in calling for a *World Without Borders*, his title for a book published in 1972.

As time went on, the argument for One World as a minimizer of the threat of nuclear destruction became enriched with an argument from social justice. The "conquest of the atom" must be made part of a global revolution that would also abolish all inequalities in the distribution of the world's resources to the world's peoples. Barbara Ward, an English economist turned sociologist, in 1962 gave an effective presentation of the new crusade:

I am deeply concerned with [this] aspect of good order: the ability of the rich to recognize their obligations and to see that in an interdependent world—and Heaven knows our interdependence cannot be denied when we stand in the shadow of atomic destruction—the principles of the general welfare cannot stop at the limits of our frontiers. It has to go forward: it has to include the whole family of man.[2]

In this view, we are all made kin not so much by being the children of the same God but by the coming universal ability to kill one another. In 1964, the Canadian journalist Marshall McLuhan developed the idea in a book in which he said that the electronic contraction of distance has made "the globe no more than a village."[3] From this none too clear assertion was derived the term *global village*, for which McLuhan was thereafter given credit. Other voices spoke of the desirability of creating an *open community*.

When you think deeply about the last two terms you realize that both are oxymorons, but they are nonetheless powerful. Above all else, a village evokes an image of something small and intimate, while the psychological essence of a true community is that it is *not* open to every wind of opinion that might batter it from the outside.

The word *village* brought small-group ethics into the discussion, while the compressive term *global* permitted the mind to ignore the demands that large numbers might make on anyone living in a totally interdependent world. By this journalistic maneuver, the popular understanding of ethics was once again freed from the embarrassing burden of scale effects. Ethical principles developed for a community of ten dozen citizens were lightly assumed to be no different from the principles needed for a world of 5 thousand million. Unmentioned was the fact that scale effects necessarily shape the default positions of theories that deal with the interactions of large numbers of people.

This sleight of mind escaped notice at the most elevated levels of the intelligentsia. On Christmas Day 1975, in Haifa, Israel, the Technion-Israel Institute of Technology managed to secure the signatures of a most distinguished body of men to the following statements: "Ultimately all must benefit from the promise of technology or all must suffer—even perish—together" and "Absolute priority should be given to the relief of human misery, the eradication of hunger and disease, the abolition of social injustice and the achievement of lasting peace."[4]

The signers were Kenneth J. Arrow, Lord Ashby, Ian Barbour, Daniel Bell, Isaiah Berlin, Max Black, Mario Bunge, Derek de Solla Price, René Dubos, John F. Edsall, Jacques Ellul, Amitai Etzioni, Dennis Gabor, Hans Jonas, Abraham Kaplan, Rollo May, Robert S. Morrison, Niko Tinbergen, C. F. von Weizsacker, George Wald, Alvin M. Weinberg, Lynn White, Jr., and R. H. Whittaker. The list boasts several Nobelists but no women. Is the omission significant? One wonders. One also recalls the aptness of the title of a 19th-century book: *The Madness of Crowds.*

Students of reform movements will note other aspects of this effort. Characteristically, some of the reformers sought to kidnap the cause, trying to warp it toward other goals—the eradication of poverty, hunger, and disease; the abolition of injustice; and (of course) the creation of universal peace. Forgotten is Voltaire's warning that "the best is the enemy of the good."

Even the most distinguished assembly is not immune to what Francis Bacon called the "idols of the theatre." (A modern equivalent is "playing to the gallery.") In the 1960s there arose a spirit of intransigence among the youth of the United States. Because of the baby boom immediately following World War II, the adolescent population was unusually large. The ever-present Oedipus process led to a justifiable disrespect for their elders' rationalization of the wars they had created in Southeast Asia. The too numerous young took to presenting "nonnegotiable demands" to their nominal superiors in colleges and universities. The Haifa statement—"All must benefit . . . or all must suffer—even perish—together"—is a typically adolescent demand made by this surprisingly old group. To label the call *utopian* is not to praise it.

"Hitch your wagon to a star" advised Ralph Waldo Emerson in the 19th century, and the dream of One World became an unusually powerful inspiration following the development of nuclear warfare in the mid-20th century. "One World or None" became a rallying cry of the disillusioned.

But what if complete union can itself lead to political destruction by a different route? As it happened, this possibility had already been suggested by a few thoughtful people long before $E = mc^2$ burst upon the political scene.

What, we must ask, if achieving One World makes its continued survival impossible? Bertrand Russell put clothes on this proposition. His argument, though brief, deserves a chapter of its own.

Russell's Theorem

Faced with occasional doubts about the dearly beloved dream of One World, the ostrich that is in all of us buried its head in the peaceful sand. This peace was momentarily broken in 1948 when the English philosopher Bertrand Russell came forth with a few well-chosen words about the Human Possible. "Always," he said, "when we pass beyond the limits of the family it is the external enemy which supplies the cohesive force" of a group larger than the family. Continuing, he wrote:

> A world state, if it were firmly established, would have no enemies to fear, and would therefore be in danger of breaking down through lack of cohesive force.[1]

Saying *"in danger of"* breaking down is too weak. The situation would be much worse and justifies saying "would be *certain* to shatter into smaller bits." Even a philosopher as courageous as Russell—or as foolhardy, some would say—was sometimes a bit of an ostrich.

The year 1948 was just about midway in the Cold War, with its developing illusion of the Communist world of tomorrow. During the period from 1919 to 1989, the Russian proletariat expressed no public doubts about this fantasy, while the timid of other nations (an association of both liberals and conservatives) had moments when they feared that Soviet Russia might be on the right path. How lovely it would be if humanity *was* on its way to evolving into a single, peaceful, worldwide sovereignty! It took the precipitous breakup of the USSR to awaken the ostriches, who had acted as if they were unaware of similar disintegrations already well advanced in central Europe and central Africa (to mention only a few of the pimples on the face of the contemporary world). The disintegration of the USSR was a wake-up call to political reality.

The current breakup of large nations into smaller units was finally recognized as a reality that could not be ignored. Russell's thesis fitted all too well with the totality of world history once observers took off their rose-colored glasses. His thesis is based on human experience of the most general sort. The Russellian thesis amounts to a default position of the sort favored in all the true sciences. It assumes that selection favors aggressiveness—at least in the short term, when wars are set under way.

By contrast, the fashionable dream of One World presupposes that aggression is nullified by the domination of altruism over egoism. But if ours is a limited world—which necessarily includes serious scarcities—a different picture emerges. The survival of reproducing units in such a world gives egoistic impulses an advantage over altruistic ones. Aggression is called forth. (We can admit that altruism does survive in a concerto of collaborations with egoistic elements, but a Johnnie-One-Note Symphony of Pure Altruism does not exist.)

One World cannot endure—not in a universe programmed by natural selection (which is the only universe we can imagine). E. B. White, the thinker-in-residence of the *New Yorker* magazine for many years, anticipated Russell when he said, "The awful truth is, a world government would lack an enemy, and that is a deficiency not to be lightly dismissed."[2] White was a great stylist; he did not employ the adjective *awful* thoughtlessly.

There is an emotional kinship between White's *awful* and Tertullian's *blessing*. That the impossibility of One World could be a blessing was acknowledged by scholars as long ago as 1860 when Goldwin Smith, a professor of modern history at Oxford, wrote:

> If all mankind were one state, with one set of customs, one literature, one code of laws, and this state became corrupted, what remedy, what redemption would there be? None, but a convulsion which would rend the frame of society to pieces, and deeply injure the moral life which society is designed to guard. Not only so, but the very idea of political improvement might be lost. . . . Nations redeem each other. They preserve for each other principles, truths, hopes, aspirations, which, committed to the keeping of one nation only, might, as frailty and error are conditions of man's being, become extinct forever.[3]

In engineering terms, a multinational world (whatever its faults) has a built-in safety factor. One World has no safety factor; it is a terrifying gamble on "one world or none." The alternative, and saving, principle was understood by Cervantes in the 17th century; in *Don Quixote*, he advised that we should never "put all our eggs in one basket." One World = One Basket, whether or not we recognize the equivalence.

Yet the fantasy of One World will no doubt continue to undermine many of our attempts to solve global problems. Time after time, events following well-meant efforts to improve the world have produced more losses than gains. Do such experiences cool the ardor for other changes? Not at all. Spokesmen hell-bent on achieving One World excuse each failure by saying, "Well, anyway, it was a step in the right direction!"

But if Russell is right—and if the idealistic last step is a fatal one—then failed intermediate steps should be regarded as steps in the wrong direction. If Russell was right—and biologists and engineers are inclined to agree with him in this conclusion—rather than fruitlessly trying to move one step closer to an unattainable goal, it is more sensible to explore the means whereby a many-partitioned world can live in a dynamic equivalent of balanced antagonisms.

As we attain this wisdom, we will no doubt find both utility and beauty in the essential nature of things. At first glance, the problems created by overpopulation are truly *awful*, in the primitive meaning of the word. Accepting this reality, we should yet be able to make the solution of the truly frightening problems of overpopulation into something of a blessing in Tertullian's sense. This blessing can be best realized if we rid ourselves of some of the illusions that have clustered around the dream of total equity. Such is the task of the next chapter.

13

A Martian View
of Malthus

I suppose I must have been fatigued by the effort of trying to pull my head out of the comforting sands of the One World delusion. Like many of my contemporaries, I had first adopted that posture during my high school days. Belief in an inevitable evolution toward a universal sovereignty was a noble fatalism of my generation, unsullied as it was by experience. Now I found that my Martian friend, the great disseminater of objectivity, had entered the room. Without giving me time to muster my defenses, he berated me.

"In the past, you spilled a great deal of ink trying to get Earthlings to take population problems seriously; but I don't recall your ever having told them, in painfully explicit detail, what they can actually *do* about the threat of overpopulation. Have you?"

"No," I replied, "if you put it that way, I guess the answer is 'No.' But how can I? No one knows what to do."

"It depends on what the word 'knows' means. You certainly could describe explicitly the roadblocks that are presently holding up progress."

"Just what do you have in mind?"

"In two words, Tertullian and Malthus!"

"But I've written a great deal about Malthus—about what is sound in his argument, as well as what is not sound."

"All taken no doubt from the *first* edition of his essay—right? Why have you ignored what came after that? Are you afraid?"

I protested. "Almost nothing came after. The first edition was a graceful essay. Then he responded to pressure from his critics and transformed his graceful essay into a boring treatise. Almost three times as many words, but no more real light was thrown on the subject."

"You forget about that great paragraph in the second edition."

"What great paragraph is that?" I asked.

"What, you too? All the world ignores the blistering passage Malthus added to the second edition, published five years after the first. I have it here: let me read it to you."

> A man who is born into a world already possessed, if he cannot get subsistence from his parents on whom he has a just demand, and if the society do not want his labour, has no claim of right to the smallest portion of food, and, in fact, has no business to be where he is. At nature's mighty feast there is no vacant cover for him. She tells him to be gone, and will quickly execute her own orders, if he does not work upon the compassion of some of her guests. If these guests get up and make room for him, other intruders immediately appear demanding the same favour. The report of a provision for all that come, fills the hall with numerous claimants. The order and harmony of the feast is disturbed, the plenty that before reigned is changed into scarcity; and the happiness of the guests is destroyed by the spectacle of misery and dependence in every part of the hall, and by the clamorous importunity of those, who are justly enraged at not finding the provision which they had been taught to expect. The guests learn too late their error, in counter-acting those strict orders to all intruders, issued by the great mistress of the feast, who, wishing that all guests should have plenty, and knowing she could not provide for unlimited

numbers, humanely refused to admit fresh comers when her table was already full.[1]

"Now," continued the Martian, "Why are you, who are so keen to have people appreciate Malthus, forever silent about this passage?"

By this time, I was nearly apoplectic. I sinned against the Holy Ghost by delivering a painfully long sermon.

▲　▼　▲　▼　▲

"At the very beginning of the Feast of Malthus it is clear that his ideal world is oriented toward the past: 'A man who is born into a world *already* possessed'—where does that leave all the new poor of the world—an increase nowadays of some 90 million each year? Malthus says they have 'no claim of right' to the existing wealth. Malthus was an ordained minister, but what sort of Christianity was he practicing?

"If a man in need cannot persuade his family to support him he must 'work on the compassion' of other guests at the feast. If they too say 'No!' the mistress of the feast (nature) says simply, 'Begone!' If, nevertheless, superfluous guests get in, 'the order and harmony of the feast is disturbed'—what a puny verb!—as 'plenty' is converted to 'scarcity.' The 'happiness' of the elite is destroyed by the 'misery' all around them. These conventional terms hardly do justice to reality. Small wonder that the outcry against this paragraph in the second edition caused their author to remove it, never to repeat it during the remaining 31 years of his life. He never reprinted it; he never defended it; he never repudiated it. *Silence.*"

As I went over this catalog of horrors, the Martian started to fidget. Finally the dissatisfaction became too much for him, and he interrupted me.

"I think your reaction to Malthus's rhetoric is too measured, too free of emotion. Let me juxtapose another reaction—mine. As it happened, I first read Malthus when I was getting acquainted with the poetry of your William Wordsworth. Now poetry writing is not our 'thing' on Mars, but we indulge in it sometimes at parties. Probably when we have had too much to drink. I don't know. But I do know that once, when my mind had become saturated with Wordsworth's sonnets, I tried my hand writing one myself, stealing shamelessly from several of his works. The result may not be very good poetry, but I think it is more honest than your serene response to the feast.[2] Let me read it to you:

TO MALTHUS

(OUT OF WORDSWORTH, *Ill Remembered, in Ill Times*)

Malthus! Thou shouldst be living in this hour:
The world hath need of thee: getting and begetting,
We soil fair Nature's bounty. Sweating
With 'dozer, spray and plough we dissipate our dower
In smart and thoughtless optimism, blocking the power
Of reason to lay out a saner setting
For reason's growth to change, adapt and flower,
In reason's way, to weave that long-sought bower
Of sweet consistency.—Great Soul! I'd rather be
Like you, logic-driven to deny the feast
To those who would, if saved, see misery increased
Throughout this tender, trembling world.
Confound ye those who set unfurled
Soft flags of good intentions, deaf to obdurate honesty!"

I had difficulty keeping my temper as I said: "Your sonnet is certainly cut from the same callous cloth as Malthus's feast. Our minister takes as fact that there are no vacant places for some of the people God has allowed to be born; that misery must increase at times; that (in a world of limits) charity is, on balance, cruel. Meanwhile, would-be population controllers complain that humans are spoilers of the natural beauty of this trembling world. You despise good intentions. So what do you want? Bad intentions?"

"You're getting warm," the Martian said. "One more step and you'll agree that the quality of an act is generally independent of the intentions behind it. Your God apparently gives credit in Heaven for good intentions; our Martian God does not. Thus it comes about that, time after time, Earthlings indulge in well-intentioned acts only to see misery increased. You particularly like to save infant lives 10,000 miles from where you live, thus increasing the suffering a generation later. But you are too near-sighted in both space and time to see what you now do 10,000 miles from home."

"But what are we to do?" I asked in anguish. "God tells us that all men are brothers and commands us to take care of the imme-diate needs of our brothers. We feel so *good* whenever we share! Is it wrong to feel good?"

"No: but I assure you, it feels even better to be *right*. The Earthly God apparently thinks that the human choice is between good and evil. Our Martian God knows that the choice is always between evils—the greater evil versus the lesser one."

"What a dreadful thought!" I exclaimed. "You won't find that detestable thought in our gospels."

"Are you sure?" My Martian smirked. "Have you thought about what Tertullian said? The implications of his remarks are two. First, every living species *must* multiply exponentially—like money earning a positive rate of interest. That means that no matter how low the rate of interest, a species multiplying unop-posed would eventually outgrow the universe available to it (which is always finite). Earthlings used to titillate themselves with dreams of one day escaping overpopulation by fleeing to the moon or Mars or some other planet. But now they know that human existence would be far more enjoyable at the South Pole or on the upper reaches of Mount Everest. Who wants to live his whole life at the top of Mount Everest?

"Second, to survive, *every* species must have enemies, and the action of the enemy in decreasing the species' numbers must eventually be greater than the species' fertility. Only that way can the tragedy of successful reproduction be kept from occur-ring."

Continuing, the Martian said: "The enemy may be starvation; it may be some biological predator; it may be some biological disease. But each species has to have at least one enemy. . . . You can put the last word in quotation marks—'enemy'—because, in a deep sense, the enemy of a species is a friend.

"Do Earthlings appreciate the Tertullian-Malthusian truth? Certainly not! Every time you are shown heartbreaking photos of starving babies in a poor and overpopulated country, you interfere with the corrective action of starvation and disease, thus ensuring that there will be even more suffering half a generation later—more, because (thanks to your 'kind-heartedness') the poverty-wracked population will have increased."

"How dreadful," I said. "Why didn't God create a world in which the rate of reproduction of a species was determined by forces internal to that species? Then happiness would not have been dependent on the intermittent action of corrective enemy influences."

"Maybe Earth's God wishes he had," said the Martian, "but even God must be the servant of reality. If many individuals within a species possessed the internal adaptiveness you crave, they would soon be displaced by the 'brothers' who retained a positive rate of increase. Biology even has a name for the phenomenon I've just described: the *competitive exclusion principle*.[3] Without it, there would soon be no living organisms. God cannot outlaw this reality."

Mustering my most resonant sermonizing voice I quoted chapter 1, verse 28 of the Goodspeed Bible:

> Be fruitful: multiply, fill the earth, and subdue it; have dominion over the fish of the sea, the birds of the air, and the domestic animals, and all the living things that crawl on the earth.

"Dare we," I asked, "ignore those commandments?"

"There's no need to be too hard on the Creator. Remember this was a first attempt. The population He commanded to multiply may have been as small as two people, or possibly a dozen or so, certainly no more than a small village—one small village diluted by the territory of a huge world.

"It would take time to realize the power of exponential growth. Unfortunately, with a population that now stands at almost 6,000 million we are still repeating the advice *intended* for perhaps no more than a dozen people. In ethics, as in all other forms of knowledge, numbers matter."

"Is there no escape from this tragedy?" I asked.

"Oh, yes," said the Martian. "We can internalize the needed enemy. We can become our own worst enemy."

Horrified, I responded: "You are talking about genocide!"

"Not exactly," came the reply. "Except in a theoretical, pain-less sense. No need for bloodshed and cruelty. All that is need-ed is a reformation of the fundamental directives of society. At the present time, Earthlings simultaneously claim two rights: (1) every child born has a right to live that cannot be taken from him or her and (2) every woman has a right to bear as many children as she wants. Taken together these two rights produce the tragic consequences of overpopulation."

My spirit rising, I said: "Then if we point this out to all women, they will internalize the needed reproductive control, will they not?"

"Good luck!" said the Martian. "As a matter of fact, some will, some won't. And remember the competitive exclusion princi-ple: if fertility varies in a population that is offered options in fertility, then as the generations succeed one another, the pronatalist elements in the population will, in time, displace the ones who conscientiously limit their fertility. You will have failed to internalize population control. (And unfortunately, some of the more competitive individuals may start thinking about vio-lent alternatives. That means that you will get genocide secon-darily.)"

"Is the situation hopeless?"

"No, but the problem will not be solved by ostriches. The sanctity of life among those born can still be maintained, *pro-vided* that the individual's right to reproduce is placed under the control of the community. Specifically, I mean the right of every woman to bear babies as often as she wants to. This is one of the many aspects of individualism that shapes and guides our soci-ety. If the ostriches can be convinced that reproduction should be a right of the community rather than of the individual, then nonviolent population control becomes possible."

I'm afraid my reply was a sarcastic one: "You mean that per-suasion is all we need to get rid of this aspect of individualism?"

"Not at all," the Martian replied. "Think of other community-controlling measures. In schools and in the home, people tell their children that robbing banks is forbidden. Does that put an end to bank robbing? Not at all. But we couple our attempts at peaceful persuasion with harsh laws that punish whatever violaters we catch. Persuasion plus legal punishment can accomplish a great deal if both approaches are well designed. Neither alone is enough.

"At the present time here on Earth you have an organization called Zero Population Growth. It is beating the drums for individual limitation of fertility. No doubt it is having some effect. But, by itself, this approach cannot possibly achieve its goal. Why? Because of the old competitive exclusion principle coupled with the general default position of psychology that 'we get what we reward for.' The ZPG organization has limited success among university audiences; it is virtually unheard of outside the learned community. Its eventual result is predictable: in the long run, it will decrease the relative number of educated people compared with the uneducated. I don't know of anyone who regards that as a desirable result. ZPG invests in failure.

"Propaganda in favor of reducing fertility must be accompanied by repressive legal measures. (There is much room for inventiveness here.) Perhaps the first thing to do would be to cancel income deductions for the third child in a family (and beyond). If that is accepted and has the desired effect, then we can think about stronger measures. Individualism need not be attacked in all its aspects; but when it comes to childbearing, we need to draw on lessons learned from other forms of community control. The community at large will be saddled longer than the parents with the burdens of the children produced. Individuals need to willingly give up some of the control of their fertility in order to benefit from an improvement in the prospects of the community—*and* to improve the prospects of their children's future.

"A sticking point in this delicate social and political transition will be people's interpretation of human equality. Let's look at that next."

Equity, Equality, and Affirmative Action

Humanity, trudging along a path presumed to lead to perpetual progress, has come to a curious bifurcation. The more familiar of the two forks proclaims that the resources available to our species are not meaningfully finite. If this is so, it may be legitimate to assume that more people are better—as are more highways, more labor-saving machinery to be kept in repair, and more information pouring in on us per unit time. A hectic life, perhaps, but a defensible one.

The other path, the ecological path, is committed to the belief that the maximum is seldom or never the optimum. If this is truly the path we should follow, problems of choice among people competing for wealth and income become serious. Choice implies *discrimination*; those who find that word alarming are apt to deny the necessity of deciding by insisting on the equality of all individuals. The ancient arguments we are then drawn into become ever more agonizing as the world's limits press ever more closely against our daily lives. In the race of life, are all people truly equal? Or, if they are not, what of it?

The rhetoric accumulated around the word *equality* is puzzling. More than 2,000 years ago, Aristotle said that the very definition of the word is equivocal. In 1840, the perceptive French traveler Alexis de Tocqueville, discussing democracy in America, wrote:

> I think that democratic communities have a natural taste for freedom; left to themselves, they will seek it, cherish it, and view any privation of it with regret. But for equality their passion is ardent, insatiable, incessant, invincible; they call for equality in freedom; and if they cannot obtain that, they still call for equality in slavery.[1]

Where do we stand today? Is it at all obvious *why* any people have a passion for equality that is "ardent, insatiable, incessant, [and] invincible"? For a social psychologist, the "why" is easily answered: *envy*, the great and silent motivator of competition, is at the base of the passion. But the subject of envy, as Helmut Schoeck has demonstrated,[2] is pretty much taboo. As a result, discussions of equality are seldom as enlightening as one would like. In pursuing his investigations into the subject, Peter Westen found that the research libraries of the United States listed the following number of titles of new books added under the category of "equality": for 1986, 50 titles; for 1987, 65; for 1988, 46; and for the decade 1978–1987, 370, making an average of about 40 new books per year.

The research worker is officially supposed to have read all the literature on a topic before publishing his or her magnum opus. How realistic is it to expect an investigator to be thoroughly grounded in "equality"? Not very; he or she will have trouble keeping up with some 40 new publications per year.

Mathematicians find equality such a simple topic that they feel little need to write books about it. But when we come to political affairs, it is apparent that the popular topic of equality is a deputy, a stalking horse for the taboo subject of envy.

The confusion is particularly acute in American thought. However variable public education may have become, it still includes one constant component: early and repeated references to the words of the Declaration of Independence, "All men are created equal." The way in which this statement is inserted into public discussions reveals that, at the very least, people think of it as a fundamental default position of political science. Some of the best minds speak to this interpretation. The philosopher Isaiah Berlin said: "equality needs no reason, only inequality does."[3] Biologists demur: at the risk of offending other specialists, they maintain that a sound theory of political science has to begin with the contrary default assertion:

No two human beings are created equal.

Multitudes of humanity, including "primitive," unlettered peoples, would not only agree with this assertion, but would be struck with amazement that anyone would hold the contrary view. Since, in our culture, it "is not nice" to point out that *equality* is a deputy for *envy*, we are reduced to asking this pragmatic question: why do Americans (and a few others) repeat a false statement over and over? What do speakers hope to achieve by building their arguments on sand?

Before delving into the origins of this fiction, a minor footnote needs to be added. Identical twins, derived from the splitting of a fertilized egg, should, barring rare somatic mutations, have identical genetic compositions and should be absolutely alike physically. Actually, however, even identical twins often have significantly different birth weights because one of the embryos benefited from a more favorable position in the uterus. Moreover, after they are born, identical twins may be subjected to observably different social environments and therefore may mature somewhat differently. Practically speaking, these qualifications are of little political importance (though they furnish an excuse for dramatists to write amusing comedies of errors).

It is quite safe to build political theory on the italicized default position stated above. To those who raise the issue of the sacred wording of the Declaration of Independence, one can only respond that apparently Jefferson had in mind something other than the equality spoken of by mathematicians. Jefferson, a Virginian, was no doubt influenced by an earlier state document that said that "every man shall be equal before the law." Such a statement is *not a theory but a political definition*. Of a definition we do not ask "Is it true?" but only "Do we accept this particular definition? Why, or why not?"

The idea of this legal definition is well captured by the classic image of blind justice. We want her to be blind to the richness or poverty of the dress of the petitioners standing before her. Our statues of her are blindfolded, but there is no reason to think that the blindfold covers her ears. Her judgments are default positions, subject to change when presented with adequate information.

"But," says the critic, "are those facts relevant? After all, under the microscope we can find no two snowflakes that are precisely alike, but that doesn't really matter. Are the differences among people just as unimportant as the differences among snowflakes?"

Eventually we will have to tackle the question of significance. Reluctant to criticize the sacrosanct Declaration, an American traditionalist is likely to fall back on the modified assertion *equal but different*. This formula verges on being an oxymoron, but it no doubt owes its popularity to the fact that *equal* permits the individual to think that "I'm as good as anybody!" while *different* implies that "I'm really something special." (Thus can envy be satisfied.) This double interpretation of *equal* is also captured in the oft-repeated assertion that "all men are equal in the eyes of God," a doctrine that Christians understandably cherish. (But does celestial equality necessarily imply equality here on earth?)

Human beings are social animals, reacting to one another. With experience the members of a group of social animals—*any* social animals—will rank themselves. Differences in power become augmented by *positive feedback* (the engineer's descriptive term) into *vicious circles* (the moralist's judgmental term). As cynics say: "The rich get richer, and the poor get children." Differences may be stabilized into social hierarchies, the interactions of which are the very stuff of history.

It can be argued that *different but equal* need not be an oxymoron. That short sentence can be thought of as referring to two different but related ideas: *equality* and *equity*. Equality is the sort of thing that the mathematician emphasizes when he or she uses an equals sign. The expression on one side of the sign can be replaced by the expression on the other side. Equity, however, implies no such equivalence, but instead is used to suggest fairness or justice. It is our desire for equity that makes us say, "One person, one vote." We know that some people are wiser than others, but we foresee too many difficulties if we try to match voting power with brainpower. That does not mean that equity is unbounded. As far as voting is concerned, we define the very young as *outside* the legal bounds of equity in voting; so also with extreme forms of mental deficiency.

However much we may treasure equity, we should not rule out the possibility that a degree of true inequality may be required for stability in any society. Snowflakes may be completely interchangeable; but a society of people, subjected to a great variety of microenvironments, is most likely to be stable if feedback and random migrations of people who are slightly different lead to assortative job placement: round pegs for round holes, square pegs for square holes. Assortative placement favors contentment and, hence, social stability.

Long before history was captured in writing, some people (the losers in life's competition?) questioned the equity of the hierarchies society had drifted into. The apostle Peter said that "God is no respecter of persons" (Acts, 34:10). Paul, in his epistle to the Galatians (3:28), said that "There is neither Jew nor Greek, there is neither bond nor free, there is neither male nor female: for ye are all one in Christ Jesus." In the second passage we see a hint of the political upheavals to come during the 19th and 20th centuries in the Europeanized fraction of the world.

It is hard for any generation to imagine the mental furniture of its ancestors more than three generations back—about a century ago, say. The respected political economist John Stuart Mill (1806–1873) said that his fellow Englishmen found "the very idea of equality strange and offensive."[4] In the same era the journalist-economist Walter Bagehot defended the English reverence for rank—especially hereditary rank. Of course, Mill and Bagehot by no means constitute a random sample of Victorian opinion since they were of the elite. But at that time, even more than now, elite opinion passed as public opinion.

Popularizers of elite Victorian opinion found a ready audience for their productions. A poem written by Cecil Frances Alexander was turned into a hymn, parts of which were still sung in some American churches in the 20th century. The third stanza of "All Things Bright and Beautiful" tries to reconcile powerless people to their situation:

> The rich man in his castle,
>   The poor man at his gate,
> God made them high or lowly,
>   And ordered their estate.

If asked for a justification of this verse, an elitist would no doubt remind others that, for the sake of social stability, pegs should be put in properly shaped holes. Even so, as the 20th century wore on, the sentiment of this poetry became irreconcilable with what was popularly held to be the American spirit, and this particular stanza was often left out of American hymnals.

Ironically, this poem came out in print in 1848, the year that saw the publication of Marx and Engels's *Communist Manifesto*, which, after asserting that "the proletarians have nothing to lose but their chains," ended with a *fortissimo* trumpet call to the lowly to seize their proper power by turning society upside down:

> Working men of all countries, unite!

▲ ▼ ▲ ▼ ▲

Failure to disentangle the concepts of equity and equality resulted in some political statements that have had fantastic effects on the course of history. In writing the Declaration of Independence in the early summer of 1776, Jefferson said: "All men are created equal and independent." (When we parrot this statement in our day, we intend that *men*, should stand for *men and women*. It is always possible for a speaker to mean by "man," the generic Latin *homo*, not the masculine *vir*. Unfortunately English makes *men* do for both meanings, thus promoting needless dissension.)

Biologists cannot passively accept Jefferson's claim as a simple statement of fact. The more extensive genetic analysis becomes, the more obvious it is that all human beings are, by reproduction, created *un*equal. The science of genetics scarcely existed until the 20th century; but even before then, the folk knowledge that guided animal and plant breeders in their work was good enough to justify scoffing at the language of the Declaration.

Jefferson was an intelligent man with rural roots; how did he happen to be so wrong? Such history as we have of the origin of the language of the Declaration makes the mystery even greater. Jefferson was personally familiar with George Mason, who had already drafted the Declaration of Rights for Virginia in the late spring of 1776, in which he wrote:

All men are born equally free and independent.

A critical biologist can accept Mason's statement precisely because it is technically true. All babies are "equally free and independent" simply because every one of them is neither free nor independent. They are equal in the fact of their dependence on their caretakers. This textual defense may be disparaged as jesuitical, but it cannot be refuted. Returning to the distinction made earlier, we can say that Mason was making a true statement about equity, which Jefferson twisted into a statement that could easily be misinterpreted as concerned with equality.

Such fine-tuning of the rhetoric may be met with a sophomoric rebuttal: "Stop quibbling: you know what I mean!" This is an unacceptable defense because too many Americans show, by the expensive litigations they launch, that they confuse equity with equality. Consider the following cases, both from Los Angeles in 1979.

1. The deaf executive director of a county Council on Deafness, after being rejected for jury service, brought suit, asking the court to compel the officials to put her on the jury. In a five-page statement the judge rejected her suit, saying that the law required that a prospective juror be in possession of his or her natural faculties.

2. The U.S. Supreme Court, in a 9-0 decision, refused to reverse a lower court that had let a school of nursing deny admittance to a deaf would-be student. Counsel for the woman pointed out that the Rehabilitation Act of 1973 requires that "no otherwise qualified handicapped individual . . . shall, solely by reason of his handicap, be excluded." The court said that a deaf person fails to meet "reasonable physical qualifications" for the job of nursing. A short time after this verdict was announced, some

150 handicapped protesters marched in Los Angeles, chanting: "Rights have no price" and, "How would you feel if it was you?"[5]

A few more comments are in order. A nonlawyer might think that the first suit was so ridiculous that it would have been thrown out with a single word, but such is the loquacity of the law that the judge took five pages to say the obvious.

In reacting to the second judgment, note that the protesters raised the issue of *rights*. The literature on rights must be at least equal in quantity to the literature on equality. Both are rich in ambiguities. Saying that "Rights have no price"—that quantities are irrelevant to justice—is unacceptable and has been since the day when Galileo said that the book of science "is written in the language of mathematics." Three centuries later the philosopher Alfred North Whitehead carried the thought forward:

> Through and through the world is infected with quantity. To talk sense, is to talk in quantities. It is no use saying that the nation is large,—How large? It is no use saying that radium is scarce,—How scarce? . . . Elegant intellects which despise the theory of quantity, are but half developed.[6]

The political implications of all this can be easily summed up: equity is determinable by law and custom; equality is determined by nature.

It is fashionable to express disgust at putting a price on the precious things of life—beauty, honor, life itself. It is true that many of the goods of life cannot easily be assigned precise numbers; but they can be compared with one another and organized into an approximate hierarchy of greatness. In the second of our two examples, we cannot both satisfy the would-be nurse's pride *and* relieve the anxiety of the patients she might later tend to. Ours is a limited world. In comparing joined alternatives, the default position should always be:

More of x means less of y.

Those who want to sell something are not enthusiastic about giving publicity to this last truism. They want the limitedness of the world to remain under a taboo; a trumpeting of the truth might inhibit sales and diminish the Gross National Product. Yet all of us, as the beneficiaries of unearned wealth, have to make do with limited resources. Choices must be made. To reject comparisons, to reject numeracy, to reject choice is to reject rationality itself. If any special-interest group insists on exclusionary rights, spokesmen for society must then ask: "What right do you have to insist that the rest of us become poorer so that you may become richer?"

The question "How would you feel if it was you?" reveals an important problem faced by the handicapped. "Equal in the eyes of the Lord" is a cry from the heart to be *accepted* by some larger community (preferably here on earth). Those who are not handicapped need to be made to feel what it is like to be rejected for deficiencies not of one's own making. But at least two communities are involved in the situation under consideration, and advocates for the smaller community—the deaf, in our example—also face a challenge to their imaginations. They need to imagine how the quality of life in the greater community would be diminished if the reform they demand was adopted. What would it be like to be a helpless patient in a hospital, dependent in the middle of the night on being heard by a deaf nurse? The example is not forced.

▲ ▼ ▲ ▼ ▲

Accepting the distinction between equity and equality permits us to understand the reason for the rise of the concept of *affirmative action* in the latter half of the 20th century. Many people are passionately affronted by what they see as cases of injustice in the placement of applicants in hierarchies: admissions to colleges, hiring of professors, awarding of the better jobs in the economy, and so on. From their point of view, affirmative action laws often give the illusion of fitting the ability of the applicants to the demands of the position. But in practice, all too often little attention is given to fit, while much is made of the rights of applicants. In legal tests, the proportion of hires is often made equal to the proportion of the identifiable group in the population as a whole. Such a hiring policy may, *by definition*, achieve equity, but there is no reason to suppose it will *in fact* produce the quality the job calls for. Let's take an extreme example. Suppose that, in the pursuit of equity, our laws mandated the admission of pygmies to professional basketball teams: would the quality of the play remain as high? Would the enjoyment of watching the game be as great? Equity and equality are often in opposition.

The passion for equity is an offshoot of the tradition of *radical individualism*. In practice, this tradition causes us to judge the rightness of an action almost solely by its direct effects on individuals claiming the right while largely ignoring its predictable effects on the larger community. As population continues to grow, it is probable that centuries-old *communitarianism* will ever more often take precedence over the more recently sanctified individualism.

The larger framework of ethics is more easily accepted by the rest of the world than it is by Americans. This point is worth remembering in our negotiations with foreign nations. Their interpretation of *rights* is not the same as ours. What *equality* means often differs significantly on the two sides of a national border. In the interests of international peace, we should be very chary of criticizing other nations for the way their people behave, in conscience, *within their own borders*.

The rhetorical dimensions of *individualism* are not the same in East and West. This point was well made by Lee Kuan Yew, for decades the guiding force of modern Singapore. Prime Minister Lee, who was educated in both Asia and England, said: "To us in Asia, an individual is an ant. To you, he's a child of God." However committed modern Asiatics may claim to be to programs of modernization or Westernization, they are much more supportive of communitarian ideals than are contemporary Westerners.

▲ ▼ ▲ ▼ ▲

Many thoughtful people are wondering if the dimensions of individualism in the West do not need to be redefined. One can argue that much of the increase in criminality is a consequence of giving too much latitude to individualism in daily life. On the other hand, it has been argued that without the unparalleled increase in individualism in the last three centuries, we would not have had the great burgeoning of science and technology. The whole world has benefited from science produced predominantly by the individualistic Western world. Those who propose altering the balance of individualism and communitarianism must ask themselves these questions: how much genuinely new science can strongly committed communitarians produce? Can you imagine a Nobel physicist coming out of an Amish colony?

The rise of radical individualism was contemporaneous with the rise of modern science. The most vigorous growth period of both began in the 17th century. Of course, *post hoc ergo propter hoc*—"after this, therefore, because of this"—is a common logical fallacy. But if we want our culture to continue to be favorable to the growth of science, we must not lightly undermine the urge to individualism and the inequalities that it sometimes fosters as we seek to enrich our way of life with more communitarian considerations.

▲ ▼ ▲ ▼ ▲

René Descartes (1596–1650) laid one of the foundation stones of science in his *Discourse on Method* (1637). As he put it, wanting "never to accept anything as true which I did not clearly and distinctly see to be so," he tried to doubt everything at the outset. Did that mean he had to doubt his own existence? He was not willing to go so far. "At least while I doubt, I must exist, and, as doubting is thinking, it is indubitable that while I think, I am." These last four words are better known in Latin (which was virtually the common tongue among the learned of Descartes's time): *Cogito, ergo sum.*

Philosophers have written many erudite pages picking the nits out of Descartes's sheltering wig, but most people—certainly most scientists—think Descartes's assumption is close enough to the truth to enable a reliable science to be built upon it.

Galileo evidently thought so. In 1657, only seven years after the death of Descartes, he joined with others in Florence to found the oldest science academy still in existence, the Accademia del Cimento. The Italian word *cimento* means "risk" or "danger"—a suitable warning to all who contemplate becoming scientists, who must often risk breaking taboos.

Many areas in need of more certain knowledge have "taboo"

signs posted at their boundaries. We note that Paul Sears has labeled his science—ecology—the "subversive science." To see how subversive of popular beliefs a new science can be, one has only to read the newspapers published since 1962, when Rachel Carson hit the public over the head with her *Silent Spring*. It took almost a full generation for the general public to admit that the insights of the newly publicized science must be admitted into the community.

15

Multiculturalism

For and Against

Evaluating different ways of life—especially the ways of one's own people compared with those of others—may seem at first glance as beyond the reach of objectivity. The distinction between *ours* and *theirs* is basic to the individual's biology; it is also basic to traditional groups, particularly when the different ways are tied to religion. The public passion for objectivity, so important in the nourishment of science and the scientific attitude, came late in the development of civilization. It was greatly strengthened by the invention of the mythical Man from Mars, whom we compel to question our prejudices.

Attaining a fruitful objectivity toward traditional rhetoric was greatly aided in 1900 by the coinage of a new word, *ethnocentrism*. Six years later, this word became a major analytical weapon in W. G. Sumner's influential book *Folkways*. Sumner defined the attitude as follows:

> *Ethnocentrism* is the technical name for this view of things
> in which one's own group is the center of everything,
> and all others are scaled and rated with reference to it.[1]

Among the common folk, as Sumner pointed out (using the

words of the poet Alexander Pope), "Whatever is, is right." These four words mirror almost the *only* attitude of the vast bulk of humanity for almost all of historic and prehistoric time. Naming the attitude, however, invited the doubting of it: in 1951 the anthropologist E. E. Evans-Pritchard wrote that the "ethnocentric attitude has to be abandoned if we are to appreciate the rich variety of human culture and social life."[2] The abandonment of traditional ethnocentrism took place first among the elite of society, principally the literati, who often lead a somewhat magical creative life in urban centers. (Think of American literary life in Paris in the early 20th century, as well as in the Greenwich Village area of Manhattan.)

Scientists, if one judges from their scarcer autobiographical writings, appear less commonly to pass through the phase of *Weltschmerz* ("world sadness") that is so characteristic of literary adolescence. On the other hand, the budding scientist's conviction that he or she is engaged in work that will ultimately be world-shaking often leads to a developmental period that can only be called euphoric. C. P. Snow, an English physicist and novelist, has recorded his heartfelt gratitude to the elite community of nuclear physicists of which he was privileged to be a member in his youth.

> . . . [U]nless one was on the scene before 1933, one hasn't known the sweetness of the scientific life. The scientific world of the twenties was as near to being a full-fledged international community as we are likely to get. . . . [T]he atmosphere of the twenties in science was filled with an air of benevolence and magnanimity which transcended the people who lived in it.
>
> Anyone who ever spent a week in Cambridge or Goettingen or Copenhagen felt it all round him. Rutherford had very human faults, but he was a great man with abounding human generosity. For him the world of science was a world that lived on a plane above the nation-state, and lived there with joy. That was at least as true of those two other great men, Niels Bohr and Franck, and some of that spirit rubbed off on to the pupils round them.[3]

The passage from which this is taken includes additional names that clearly come from many national cultures. Copenhagen was a truly intoxicating mixture of investigators united by their love of science.

To be a "citizen of the world" was the proud claim of many scholars whom we now call philosophers, from Socrates (5th century B.C.) to Zeno (4th century B.C.) to Francis Bacon (17th century A.D.). From this time on, the claim was made by more and more natural scientists, and with good reason: the truth they were seeking was clearly the same under all jurisdictions, whereas the truths of ethics and politics seemed to vary with the culture. As science grew in prestige, so did the passion of non-scientists for international orientations. Multiculturalism, born among the elite (whether literary or scientific), was eventually confiscated by the much more numerous proletariat.

A distinction can be made between two forms of science, the names of which are various. What are now generally called the *natural sciences* include physics, chemistry, and biology. In these sciences, speaking the truth does not falsify it. A true characterization of gravity does not set in train a series of events that produces, say, a repulsion. When our predictions fail to come true, it is because we have made simple errors of some sort (e.g., $2 + 20 = 42$).

The contrasting area is that of the *social sciences,* also called the *behavioral sciences*: sociology, political science, and so on. Statements in these sciences are often falsified by the words we choose to describe the facts. The assertion of the poet Robert Burns that "The best laid schemes o' mice and men gang aft a-gley" is also true when social observations become public knowledge. (A comparable phenomenon in the natural sciences would be for the act of describing gravity to convert it to levity.) In the behavioral sciences, the vulnerability of "laws" to the effects of publicity creates serious problems with regard to responsibility, free speech, and censorship.

So: clear verbalization of a law in the behavioral area may induce a contrary law. Thus it happened that a half century's derogatory remarks about ethnocentrism generated a new offense of the opposite sort, namely, *ethnofugalism* (from the Latin *fuga,* meaning "flight")—a habitual assumption that the folkways of *our* culture are absurd compared with those of other cultures. This attitude leads to a fleeing from the way we were brought up. At midcentury, fugalistic doubts were largely restricted to the elite of our society. Then, in the 1960s, disillusion with the Vietnam War and the political establishment that supported it infected the masses with ethnofugalism. Whatever improvements fugalism may produce, they do not include an increase in political stability. (A nation that wants to survive in competition with others should keep this fact in mind.)

Out of ethnofugalism grew *multiculturalism,* a practice called a virtue by its devotees. It presumes the peaceful coexistence of many cultures within the boundaries of a single political unit (usually a nation). On its face, multiculturalism suggests tolerance, an opening up of closed minds to new ideas. Says the journalist Richard Bernstein in his *Dictatorship of Virtue*: "Culture is powerfully conservative. [It] enforces obedience to authority, the authority of parents, of history, of custom, of superstition." But the fashionable multiculturalism of our time tacitly assigns a contrary quality to culture. Multiculturalists, in effect, urge that we eat borscht with chopsticks.

At a more serious level, today's multiculturalism, Bernstein says,

> is a movement to the left, emerging from the counterculture of the 1960s. . . . [It] cannot be taken at face value, and that is what makes it so tricky. Nobody wants to appear to be against multiculturalism. Hence, the irresistible temptation of the post-1960s, radical-left inhabitants of a political dreamland to use the term "multiculturalism" as a defense against exposure or criticism . . . multiculturalism has an almost salacious attraction. . . . To put matters bluntly: the multiculturalist rhetoric has the rest of us on the run, unable to respond for fear of being branded unicultural, or racist. . . . In such a way does multiculturalism limit discussion; it makes people afraid to say what they think and feel. . . .[4]

Chaos generates violence. Recall what happened during the last two decades of the 20th century in the multicultural USSR, which (curiously) had as its completed lifetime the biblical threescore and ten years (from 1919 to 1989). At about the same time, the social consequences of multiculturalism put an end to unity in the Baltic region of Europe and in central Africa. In light of these observations, it takes the moral blindness of the mythical ostrich to be a promoter of multiculturalism within a nation. Most people are better off clinging to the relative narrowness of a traditional culture while enjoying only vicariously— in the international sphere—the exceptional experiences of talented artists and brilliant scientists who move so easily between cultures. In the final reckoning, ethnocentrism is not all bad.

Future histories of the United States will have to account, in detail, for the rise of enthusiasm for internal multiculturalism. There has been a puzzling reversal of the justification of immigration expressed during the last two centuries. The American ideal used to be the *assimilation* of the incoming migrants from other nations: the word *United* implied united in culture, in ideals, in values. Without embarrassment, established residents intent on educating our immigrants to the American way of doing things would promote the Americanization of the new citizens. (Now *Americanization* is not even used to damn it: the sand of shame makes ethnofugalistic ostriches choke on it.)

The ready adoption of multiculturalism as an ideal springs from a serious misunderstanding of the nature of *culture*. This word is more than five centuries old and has undergone many mutations of meaning. In what follows we shall accept the spin given it by anthropologists in the late 19th century. The definition of *culture* presented here is one of many recorded in the *Oxford English Dictionary*: "The activities of a society—that is, of its members—constitute its culture." Accepting this broad emphasis, it is safe to say that anthropologists intended to enlarge thinkers' sympathies for other people's ways of doing things. What has been insufficiently appreciated is this: the consequences of multiculturalism are very different when the "multi-" is found *within* a single nation rather than in the variety *between* nations.

The legal profession has recently embraced *intra*national multiculturalism as a useful fiction to use when defending foreign-born clients who have violated American laws by continuing *here* with practices that are approved of in their *cultures of origin*.

Reporting—and supporting—this development, an anonymous writer in the *Harvard Law Review* in 1986 stated: "Immolating one's own children for the sake of honor, executing an adulterous wife, and lashing out at someone in order to break a voodoo spell may seem very bizarre—indeed barbaric and disturbing—to the majority [of Americans]. But this is no reason to attempt immediately to quash the values of foreign cultures. American society has thrived on tolerance, curiosity toward the unknown, and experimentation with new ideas."[5]

This argument, called the *cultural defense*, gives the legal profession one more excuse for its existence. With the cultural defense the courts are offered "an opportunity to strike the necessary balance among . . . competing interests." The question for the body politic is this: should we permit criminal lawyers to use this new weapon? Admittedly, Americans commonly regard tolerance as a virtue, but variations in virtue *within* a single sovereignty have quite different consequences from variations *among* cultures. Are not stability and predictability of the law also virtues? Does not the very word *law* imply limits on diversity? To elicit dispassionate discussion (if this is possible) we need to lay bare our principles by focusing on a cultural difference that is devoid of ethical content. Fortunately, there is one.

▲ ▼ ▲ ▼ ▲

Suppose that, among the many automobile drivers who travel by way of the Chunnel from France to England, there is a strongly principled American who, ignoring all signs, insists on driving on the right side of the road when he gets to England. The inevitable smashup occurs. After an unplanned hospital stay he is brought before an English court, where he, like Zeno of Cytium before him, proudly claims to be a citizen of the world. He rejects the parochial rules of the island on which he now finds himself.

"American law," he argues, "was derived from English law, and both systems assign the highest priority to individual freedom. Both Americans and Englishmen are free to say what they think. We should also be free to act as we wish, whenever the act creates no *intrinsic* danger. Intrinsically, there is no difference between driving on the left and driving on the right. From which it follows that no one is responsible for this accident. It's God's fault."

Put another way, there is no ethical meaning to driving on one side of the road rather than the other. In evaluating competing cultural elements, we must look at the consequences of allowing contradictory elements on the same playing field. In devising principles to guide people living together, only consequential ethics makes sense. The car driver must ask, *If I choose to drive a certain way, what will the consequences be?* The answer to this question has to be a group answer. No matter how enamored I am of philosophical individualism, I cannot be allowed freedom to act as an individual who is, in all situations, utterly unbound by conventions.

The example may be a bit silly, but the conclusion is not. Which side I drive on is an element of culture. Whenever a nation commits itself to internal multiculturalism, it is headed for trouble.

There is a necessary element of intolerance in cultural traits. For instance, try as they may, not many American enthusiasts of multiculturalism actually welcome the introduction into our nation of the practice of female circumcision by immigrants from central Africa. This procedure, more accurately called *clitoridectomy*, is carried out on a girl of 10 without the blessed use of anesthesia. In both practice and consequences it is a painful operation; nonetheless, it is supported by religious sanctions. (Though an estimated 50–100 million females have been so treated, we had better question our passion for religious freedom.) A woman who was subjected to the brutal operation as a child is apt to dread sexual intercourse ever after. Africans defend the practice because they believe it diminishes the probability of marital unfaithfulness later. Whether this belief is justified or not, do we want our children to grow up in a neighborhood in which children of African descent are subjected to this painful practice? Will not the freedom to practice this cultural trait act as a divisive force in such communities? Do multiculturalists believe that our own children will grow into nobler adults if they are thrown together with such different playmates during the sensitive years of their maturation? Do we want to tolerate torture in our midst?

▲ ▼ ▲ ▼ ▲

Indigenous cultures exhibit an unquestioning deference to traditional behavior. The commitment is tacit; challenging it often elicits passionate denunciations of other cultures. Traditionalists traditionally belittle the "other." Elitists are sometimes content to dismiss the beliefs of others as "prejudice" without insisting that they change their behavior. A separation of 10,000 miles between incompatible practices may be tolerable; 10 feet is another matter. Many idealistic people today seek more openness in human relations. But, as Fred Siegel has pointed out, an *open community* is an oxymoron: such an imagined community would destroy the emotional underpinning that is so precious to persons who feel that they are supported by *their* communities.[6] To the extent that a community *is* open, it is not so much supportive as it is indifferent. Ask anyone who lives in Manhattan.

▲ ▼ ▲ ▼ ▲

In 1936 the sociologist Robert K. Merton opened up a new line of investigation when he called attention to the *unanticipated consequences* of almost any reform.[7] In 1996 this insight was made the leitmotif of a stimulating book by Edward Tenner, *Why Things Bite Back: Technology and the Revenge of Unintended Consequences.* As a well-known example, consider the discovery of penicillin. In the early days we thought this antibiotic would completely wipe out certain kinds of disease; then we found that exposure to the antibiotic selected for genetically resistant hereditary variants of the microbes. We learned to curb our use of penicillin and to augment this antibiotic with others. Such has been the history of many antibiotics.

It is not surprising that this phenomenon should be widespread in the real world: the assumption of hereditary differences is a major default position of ecology. Less than ecological analysis assumes *one cause* ⟶ *one effect forever.* Small wonder that things "bite back" so often! Ecologists who advise on matters of disease and weed control warn the community that: *we can never do merely one thing.* Understandably, commercial operators, whose wealth is derived from ignoring the complexities of the real world, fear and detest ecologists.

Social reformers also can grow to despise ecologists, and for the same reason: the conceptual worlds of reformers are generally so much simpler than the real world that the achieved reform sometimes "bites back" at the reformers. When late-20th-century idealists sought to erase economic differences among citizens, a split in their ranks soon developed: one group sought *equality of opportunity*, while the other yearned for *equality of outcome*. The second group criticized the first for chasing after the unattainable (e.g., total control by the community to ensure that no parents gave their children a better start in life with a richer family life).

The first group was loath to tolerate present differences that traced back to earlier conditions (e.g., slavery). This group believed that people are indeed equal at the moment of creation (when the egg is fertilized by a spermatozoon). This view, however, is contrary to all of biology.

One kind of reformer is disturbed by inherited injustice; the other is terrified of "spoiling" today's children with benefits they have not earned for themselves. There is some justice—and some error—in both positions.

By the second half of the 20th century, *outcome equalitarians* had won many legal battles. A new battle cry was being heard throughout the land: *affirmative action*, which was discussed in chapter 14. The term comes from a law dating back to the year 1935, but the chain of political consequences really started with the Civil Rights Act of 1964.

As Americans became more keenly aware of the unfair treatment blacks were receiving in the labor market, they sought greater justice through law. Honestly administered tests for relevant competence yielded a proportion of qualifying blacks distinctly below their proportion in the general population. Being unwilling to admit error in the contention of the Declaration of Independence that "all men are created equal," reformers asserted that the results of the tests were explained by observable inferiority in the education of the blacks (taking *education* in the broad sense to include all family and community experiences). A regard for justice led to the recommendation that affirmative steps be taken to ensure that blacks had a fairer chance in the job market. In other words, affirmative action implies that *if we cannot guarantee equality, then we should legislate equity.*

Americans have long accepted this maneuver when it comes to voting, but should it be extended to fitting round pegs into square holes in the workplace? Some reformers proposed that each racial group be assigned a quota equal to its frequency in the general population. Since about 12 percent of the population is black, the presumption was that this should be the proportion of blacks in all occupations—including mathematics and basketball. (Interestingly enough, no public outcry developed for increasing the proportion of white professional basketball players. Such inequities led Nathan Glazer to conclude that *Affirmative Action* really should be called *Affirmative Discrimination.*)[8]

A legal test of affirmative action reached the U.S. Supreme Court in the case of Alan Bakke, who applied for admission to the medical school of the University of California, Davis.[9] His lawyers pointed out that his score on the admissions test was distinctly higher than those of a number of black students who were admitted to the medical school, while he was denied admission for two years. What principle should rule—individualism, or egalitarianism? Like Solomon, the Court equivocated: they said that affirmative action was not unconstitutional but that Bakke should be admitted.

Such is the lethargy of the law that by the time the judgment was handed down it was moot as far as Mr. Bakke was concerned: discouraged, he had abandoned his ambition to be a doctor and embarked on another path. Since 1978 many similar cases have been handled by the courts in much the same way. Our society is not sure whether it wants to take the individualistic road or the group road to justice. Equality or equity? Stay tuned.

Following the line of thought begun by Merton and continued by Tenner, we can view this as just one more example of "things biting back." Or we might take a more moralistic stance and say that experiences of this sort show the danger of investing too much emotion in hate. Hating racism, reformers created affirmative action, which turned out to be racism with a different name—discrimination with the opposite sign. And now the new discrimination is leading to bitter feelings among a new group, which is being judged as a group and not as individuals. We wonder: should we hang this needlework motto on our walls?

We become what we hate.

▲ ▼ ▲ ▼ ▲

Multiculturalism can be viewed as a sort of moral promiscuity. Those who yearn for the life of a village—of a warm community—should realize that *in the absence of exclusion, inclusion has little meaning.* Moreover, it is the feeling of inclusion that attracts people to the dream of a village. Seekers of brotherhood should not forget Proudhon's cautionary insight: "If all the world is my brother, then I have no brother."

Promoting multiculturalism within the bounds of a single country means encouraging the grossest sort of promiscuity, blind to the fact that the unity and psychic strength of a nation depend, in large measure, on maintaining the dominance of many law-supported discriminations. For the sake of world peace, spatially separated cultures must be allowed to obey different ethical mandates; but a nation that chooses to forbid the genital laceration of young girls or the driving of cars on the left side of the road cannot tolerate internal diversity in these matters. Nations that are unwisely tolerant will surely learn that multiculturalism is an oxymoron.

The world of the future will be a hybrid between One World and Many Worlds: a sort of cellular universe in which the unit cells are nations or something like them. The cellular arrangement has worked well in the human body; there is a chance that it can also work in the political body of the world, provided we have reasonable expectations of what multiculturalism can do. *Intra*national multiculturalism creates chaos and destroys national peace. By contrast, given the persistence of cultural bulwarks between sovereign nations, *inter*national multiculturalism can both promote peace in the short term and allow us to learn more about human nature in the long run. The "multi" must be spread out among many well-separated national unicultures.

Curiously, two different groups of reformers are now unwittingly working against worldwide peace. At the same time that one group of Americans is calling for an increase in internal multiculturalism, another group wants to bend all the rest of the world to our way of thinking about all ethical matters—insisting, for instance, that we coerce China into abandoning some of its methods of population control. Such coercion can hardly be peaceful for long. Multiculturalism can be a vice within a nation but a virtue in the total basket of national practices in a peacefully *inter*national world.

Ambivalent Value of Growth

A short generation after writing his celebrated *Essay*, Malthus once more tried to get a drowsy world to understand the fantastic potential of exponential growth. In his 1820 textbook of economics he wrote:

> . . . [I]f any person will take the trouble to make the calculation, he will see that if the necessaries of life could be obtained and distributed without limit, and the number of people could be doubled every twenty-five years, the population which might have been produced from a single pair since the Christian era, would have been sufficient, not only to fill the earth quite full of people, so that four should stand in every square yard, but to fill all the planets of our solar system in the same way, and not only them, but all the planets revolving round the stars which are visible to the naked eye, supposing each of them to be a sun, and to have as many planets belonging to it as our sun has.[1]

After that nonstop sentence (so typical of the century in which Malthus was born), the author concludes: "Under this law of population, which, excessive as it may appear to be when stated in this way is, I firmly believe, best suited to the nature and situation of man, it is quite obvious that some limit to the production of food, or other of the necessities of life, must exist."

Thus did Malthus open the door to a discussion of limiting factors to growth, which were later specified in detail by Justus von Liebig (1803–1873), beginning in the middle of the century. A great pioneer in chemistry, Liebig was virtually the father of scientific agriculture. *Liebig's law of the minimum* can be stated in several ways, one of the simpler ones being this: "The growth of a species is limited by whatever required nutrient is least available." To get very far, many *ifs, ands,* and *buts* have to be added to this statement, but the essence of the law is apparent in this simple form. In much of land agriculture, nitrogen is the limiting factor. In the open oceans, phosphorus is the most common key shortage. Pouring in quantities of nonlimiting factors is usually a waste of effort and money.

Space is not, in any simple sense, a limiting factor for human populations: the present population of the world could fit, standing up, on only half the area of our smallest state, Rhode Island. It takes no great intellect to appreciate that supplying standing room only is not enough to maintain a happy population. In a real sense, each American "occupies" about 9 acres of land if you include provision for highways, houses, factories, crops, and recreational areas. Even this expanse (which must seem large to a Manhattanite) does not allow for our symbolic occupancy of space in the foreign nations with which we trade.

How Many People Can the Earth Support? is the title of a recent and very carefully written book[2]—and the oft-repeated question of many essays during the last two centuries. If you fail to specify the limiting factor you have in mind, it is a silly question to ask. *Food?* What agricultural advances do you foresee? *Using vegetable food only?* If meat is ruled out of the diet, 5 to 10 times as many people can be supported by photosynthesis. *On solar energy?* Efficiently captured energy will support a much larger population. A still larger population can be supported if space heating and cooling is prohibited, as clothes are used to keep individuals warm. (As for cooling, let 'em sweat!) *Is solar energy to be augmented with fossil fuels?* That's living high on the hog (a perilous policy that we are following now—but not for much longer!).

The physicist J. H. Fremlin has taken the mathematician Karl Jacobi's advice and inverted the energy question.[3] Instead of asking, "Where can we get enough energy?" Fremlin asks, "What must we do to get rid of all the degraded energy produced by the biochemical processes of living?" Every bit of energy used eventually ends up as waste heat, which must be eliminated. While not claiming to have a precise answer, Fremlin calculates that when the earth's population grows to about a billion billion people (10^{18}), the massed metabolism required for life will extinguish human life itself—by a global fever, so to speak.

For a homely model of the suicide that humanity can choose (if it so wishes), inoculate a 1,000-gallon vat of grape juice with yeast and let it metabolize away *without any cooling system surrounding it.* The silly yeast cells will commit suicide. Earth has only radiation into outer space to get rid of its unwanted energy. (Optimists who boast that humanity need never worry about a shortage of energy must disprove Fremlin's point that a long-age of useless energy will ultimately destroy all of us if we do not abandon the religion of Growth Forever.)

The carrying capacity of a pasture for cattle is fairly easy to determine. Seeking the earth's carrying capacity for human beings activates profound questions of value, questions that the discipline of economics has, ostrich like, been evading for more than two centuries. A thoughtful cow (if there be such) might well decide that it need worry about only 2 limiting factors: grass and water. Thoughtful human beings (and there are such) can easily identify hundreds of such factors; we must declare which one(s) should be limiting.

An important thread running through the history of science is the intermittent conversion of questions couched only in ordinary words into ones using the imagery of mathematics. Asking "How many people . . . ?" is an invitation to seek a mathematical function to which we can apply the differential calculus as we seek its maximum point. Reverting to words, this implies that we want to *maximize the number of people* living in a sovereign nation (or perhaps in the entire world). A religious fundamentalist might argue that the advice (in Genesis 1:28) to "be fruitful and multiply" is a commandment to maximize the size of the human population. But the same verse tells us also to "have dominion over . . . every living thing that moveth upon the earth." The words *to have dominion over* do not translate into the single, brutal word *extinguish*. Yet that would be the consequence of an ill-conceived ambition to replace all other living things by *Homo sapiens*. And because of the interconnectedness of nature, we would find that many other things that normal people take pleasure in would also disappear as we maximized the quantity of human flesh. We must remember that the words of Genesis were addressed to a very small population of human beings who could hardly imagine the consequences of creating a world population of billions of people greedy for many more "goods" than mere human existence.

Greed is not one of the seven deadly sins acknowledged by medieval Christianity. But surely it is the more inclusive one of which the classical gluttony is merely a particular example. Seeking the maximum is a habitual practice of those who are incorrigibly greedy. Before concluding our search for answers to the population question, we must explore some of the consequences of excessive greed.

17

The Extended Reach
of Gresham's Law

Puzzling over repetitious studies of the world's population, professionals are astounded at the evasive maneuvers of "pragmatic" politicians. The Man from Mars would surely conclude that the attempts to maximize the human population must be motivated by a form of greed. But if our heavenly visitor wanted a welcome to continue his studies, he would keep this conclusion to himself.

In the desperate search for a solution to the human population problem, *space* was for a while the great escape from rationality: we would just ship off the earth's increase in population—some 230,000 people per day—and thus achieve a zero population growth rate. The simplest numerate analysis shows the witlessness of such a proposal. By the time we factor in the minimum transit time to the planets of other suns (several hundred years, at least) and the energy needed for the adventure, all except the most fanatical technological optimists are willing to admit that overpopulation would ruin the earth long before scientists invented the technology required for our species to colonize space.

Alternatively, some of the optimists thought that a demographic dictatorship would be the answer: married couples would be brutally forced to have the right number of children. But that dream also is a fatuous fantasy because there are limits even to the powers of totalitarianism. Have we, then, painted ourselves into a corner?

Fortunately, a resolution of the dilemma may be at hand. To understand it, we must first explore some well-known consequences of uncontrolled greed in various arenas of human action. We must examine a basket of controls that answer to a law that has been known for more than 2,000 years. This law is traditionally expressed with a specificity that prevents our seeing its generality. I refer to *Gresham's law*. So much has its power been missed that it is now often omitted from economics texts. When given, it is usually in a folksy form, to wit: *Bad money drives out good.*

Long before writing was invented, some greedy rogue turned out the first counterfeit coin. Anyone who becomes the innocent receiver of counterfeit money faces a moral problem. In the 1st century B.C., Cicero posed this question:

> If a wise man accepts counterfeit money in place of genuine money without noticing it, should he spend it when he realizes it is counterfeit? . . . Diogenes says yes, Antipater says no; and I rather agree with Antipater.[1]

We must praise Cicero for the rectitude of his choice, but the tentativeness of his "rather agree" shows that his understanding of the ethical problem was not deep. We need to probe more deeply.

In the 14th century A.D., the pioneer economist Oresme (or perhaps one of his commentators) understood Gresham's law, though it was not given this name until the 19th century—in honor of a merchant who had died three centuries earlier! The convoluted heritage of the name rightly hints at the failure of the learned world to appreciate the wider significance of the phenomena involved.

If people were not so much addicted to ostrichism, they would long ago have realized that Gresham's law governs more than coinage: it defines where, under total freedom in a world of limits, the payoff lies. We examine a brief catalog of its major applications.

LABOR. Laborers, when given complete freedom of movement, are subject to a redistribution process. When a businessperson can choose workers, he or she picks cheap labor over expensive labor (other things being equal, of course). It would be as foolish for a company agent to preferentially hire expensive labor as it would be for a seller to prefer being paid in counterfeit money. In both situations the usual decisions amount to investing in success (for the decision maker). If you were Einstein's God, would you invest in failure? Nature doesn't.

One could hardly maintain that it would be natural for laborers who find themselves unemployed to die peacefully and disappear. During the early phases of the Industrial Revolution in England, this seems to have been the expectation of many of the bosses. Plainly, the enterprisers were so remote from everyday reality, and so lacking in empathic imagination, that they expected the problems of unemployment to be resolved in this heartless way.

On the other hand, when society regards this degree of rationality as excessive, it may support the unemployed and their families with welfare payments. Commonized welfare then raises everyone's taxes. (This arrangement is often called *free enterprise*—which indeed it is for the richest enterprisers, who thus enjoy an almost free subsidy by the taxpayers. This tragedy of the commons continues in our day, and the people subsidized still complain of taxes.)

FREE TRADE. A nation may activate a Gresham effect by permitting the tariff-free entry of goods produced in foreign countries where the wages are lower. This setup is called *free trade*. (Note how often the public gives knee-jerk approval to the adjective *free*—a fact well known to all public relations experts.) In free trade, importers and others who handle foreign commodities gain extra wealth at the expense of their fellow citizens. The resultant unemployed laborers may be excused for verbalizing the result thus: *Free trade drives out fair trade.*

One does not expect members of the elite classes to accept this wording: instead they argue that labor competition motivates laborers to work harder, while the technologists in the country adversely affected furiously invent improvements that will lower the costs of production. (There is some truth in these two rationalizations, of course. But note: the elite still complain of being saddled with higher taxes to keep domestic laborers and their families from being impoverished by imports.)[2]

IMMIGRATION. The same goal—lowering the cost of goods produced—may be achieved by making the borders of a country permeable to the passage of human bodies—in a word, *open.* In this case, the net movement of labor will be from the low-wage country to the high-wage country. Result: a lowering of the high wages that were formerly paid to the longtime natives of the richer country. The application of Gresham logic to immigration can be summarized thus: *Immigrant labor pauperizes domestic labor.*

Open doors increase the freedom of foreign labor while diminishing that of domestic labor. Publicity virtuosi created the fiction of free trade. A handbook of industrial semantics would, of course, under the entry *seduction,* discuss both open trade and free trade. (Open *to what?* Free *for whom?* The good citizen is, by definition, an ostrich who never questions these sacred terms.)

For more than three decades, public opinion polls have shown that most Americans are opposed to the present high level of immigration.[3] The majority ranges from 55 to 85 percent, depending on how the question is worded and where it is asked. Verbal acceptance of immigration is highest among groups that are most at home in the English language; masters of the language are least likely to be displaced by immigrants who do not speak the language well. For work in the media (newspapers, magazines, radio, television), two or more generations of residence are usually required to master language to the extent required by this vocation. Few fresh immigrants can achieve this mastery in their lifetime. Mediamasters can afford to be generous, for (as Mark Twain said) "It's easy to bear adversity—another man's, I mean." And apparently it is easier for the media to photograph scenes that suggest the suffering of new immigrants than it is to prove, photographically, that widely dispersed job losses have a common cause. It takes an unusually sensitive imagination for a mediamaster to wake up to the truth that is kept at a distance from his or her daily life.

▲ ▼ ▲ ▼ ▲

Why did scholars fail to see the generality of Gresham's law? At least two elements entered into this failure—one operational, the other moral.

OPERATIONAL CONSIDERATIONS. Saying that bad money wins out in free competition does not harmonize with our optimistic view that good is more powerful than bad. This subconscious assumption saturates the education of the children of our time: it seems wrong to admit that bad ever wins over good. Are we faced with an error in semantics? Money is a medium of exchange. If less wealth is lost in an exchange, money is *more efficient*, is it not? Should Gresham's law have been worded thus: "Efficient money displaces inefficient money"?

Following the same pattern we could, with equal justice, reword the three generalizations cited above in this way that glorifies winning over principle:

> *Immigrant labor pauperizes domestic labor.*
> *Cheap foreign labor drives out life-supporting local labor.*
> *Free trade drives out responsible trade.*

The revised wordings put the winning elements in the above generalizations in the same logical basket as the idea of natural selection: they identify what we mean by *fitness* and *superiority*. If you wish, you can belittle the result as an implicit tautology. ("A is A" or "What is selected is superior" or "More efficient is better.")[4] Had some such revised wording been selected, Gresham's law would probably not have been relegated to the footnotes of economics texts.

MORAL CONSIDERATIONS. It is precisely because *bad* implies an ethical judgment that it found favor with practical people who had to deal daily with money matters. A counterfeiter can always justify his or her occupation as a practice of laissez-faire economics. But surely, as soon as money was invented, the generality of humankind must have realized (even if only dimly) the following fundamental truth: *in establishing a money system, a social organization that hopes to survive cannot tolerate absolute freedom.*

To repeat in slightly different words: individual freedom in the choice of tokens used in financial exchanges would generate a ruinous policy. The reliability of money is essential to the progress of enterprise in developing a wealthy state. In the arrangements of nature, *freedom* is relegated to an operational position that is secondary in importance to *survival.* In human affairs, there are advantages to be gained by being honest and explicit about the relative importance of these two values.

In a competitive world of limited resources, total freedom of individual action is intolerable. Hence the need for community-sensitive restrictions, ideally produced by a policy of "mutual coercion, mutually agreed upon." It is understood, of course, that mutual agreement has to be satisfied with something less than unanimity. How this plays out in the vexing question of population growth is the subject of the next (and last) chapter.

18

Summary

Can Our Ostriches Find the Will?

My plane was late getting to the New Jersey airport, so I missed my connection to California. A bad miss: now I had to wait several hours before I could catch a night flight. Having nothing else to do, I wandered into a snack bar and ordered a cup of soup. Dismally lit, the room was almost empty. I took a stool next to two sturdy young men who looked like stevedores. (There was a marine port nearby.) I am an incurable eavesdropper; it wasn't long before the conversation convinced me that I had guessed the occupation of the men correctly. Once again I was amazed at how uninhibited many Americans are in discussing private matters in public.

"Well, see," said the huskier one, "here were all these cartons that were not too big to handle. No one was around looking after things, and the stencils on the cartons said they were from a German camera company. Stupid! I didn't see why I shouldn't take what I could haul away in my little pickup, and so I did. After all, they didn't belong to anybody. The shipper would just hit the insurance company for the loss, and the receiver would be none the worse off. So I took 'em."

My childhood training should have made me look for a guard

on whom to unload this conversation. But what chance was there of getting the law enforced in this way? What had happened was in the past; my evidence was only hearsay. How long would it take to move the law? And would my family be merely amused at my missing another plane home? Would I later receive a subpoena to come back here for a trial? I didn't know much about the law, but I couldn't see myself nobly acting as the law's self-sacrificing agent.

The stevedore's action was not exceptional. In fact, until we feel in our bones how natural and common his behavior is, we cannot understand much of politics, commerce, and distributional problems—including, as we shall soon see, the recalcitrant problem of runaway population growth.

In its inception, the moral impulse is based on a person-to-person orientation. The biblical command "Do unto others as you would have others do unto you" introduces a subtle ambiguity with the word *others*. The first *others* suggests a succession of one-on-one interactions between individuals; the second *others* evokes images of armies of people linked into impersonal institutions. It is naturally hard to take seriously a command to "Do now unto a large insurance corporation as you would have an individual person do unto you at some time in the future." Yet that is the application of the Golden Rule to situations that have become increasingly common with the passage of time and the growth of population. The thieving stevedore probably truly felt that the cartons "didn't belong to anybody."

But if the stevedore's actions were criminal, mine were at least immoral. One is not supposed to stand idly by while someone else commits or confesses to a crime. As a citizen, I had an obligation to inform the authorities of a legal transgression. But I failed—on this occasion and (to tell the truth) on numerous others. So have legions of other human beings. Our reprehensible behavior has ancient roots that we must understand if we are to approach population problems productively.

The history of humanity's treatment of access to the wealth of the oceans can serve as an example of a reigning distribution philosophy. From the earliest days the ocean had open access; lacking property, it was an unmanaged commons. Tradition was solidified in 1625 when the Dutchman Hugo Grotius verbalized it in his *De Jure Belli ac Pacis*: "The extent of the ocean is in fact so great that it suffices for any possible use on the part of all peoples for drawing water, for fishing, for sailing." The assertion of an effective infinity justified the development of the economics-without-limits whose dangers we have been exploring.

By the second half of the 20th century the potential dangers of limitless access were being realized. One population of fish after another became seriously depleted. The seriousness of the problem was masked somewhat by changes in people's diet; species that were once labeled "trash fish" were found to be not so bad after all, and commercial fishermen brought them to market. Large numbers meant low prices until heavier fishing started the trash fish on the road to depletion. The Catholic Church had for centuries (in defiance of biology) decreed that fish were not meat and could therefore be eaten on ceremonially meatless Fridays. The ruling showed compassion for the hardworking poor who felt that they needed meat every day. But as the price of fish rose, compassion was extended. In the mid-20th century the Church reversed itself and said that any sort of meat could be eaten on Fridays.

Some fish populations were depleted to the point at which governments had to restrict, or even forbid, fishing in certain areas. The restrictions bore hard on particular human populations. The Newfoundland cod fishery included many workers whose families had known no other occupation for 200 years or more, and the bleak land area of the province offered scarcely any other occupation. The government asserted national sovereignty over the adjacent ocean area and then forbad even its own people to fish in it. Some 25,000 fishermen and processing-plant employees were thrown out of work in Newfoundland. The rest of the Canadian population was unwilling to let the unfortunates starve, so the government put them on the dole.

Let's look at this situation as a problem in the distribution of wealth. The unfenced ocean is a commons for the whole world. Fishermen who tap this wealth are turning common wealth into personal wealth as they sell the fish. For thousands of years no one objected to this; fishing is hard work, and landsmen had no need to fret about the removal of part of this common wealth because biological processes would soon replace it. The price of the fish could be thought of as a just recompense for the hard work of the fishermen.

But when the dole was set in place, the psychological landscape changed. When supplying a need is regarded as a group imperative, those living on the dole draw from the common wealth even as they contribute little or nothing to it. Whenever costs and contributions are so disjoined, the system cannot be said to be a responsible one. Leaving abusive words to one side, one notes that an irresponsible system of supporting human beings will tend to maximize the number of human lives as it reduces the quality of life. Does that statement seem too extreme? Then think of the psychological situation of the children of the dole-supported families of Newfoundland: with the fishing prohibition in place, they see a future without a believable prospect of a responsible job as an adult. Experiences elsewhere show that the welfare status has a strong hereditary element—hereditary in a social sense, even if the genetic component is zero. A subsidy can be both kindhearted (in its intentions) and terribly destructive (in its consequences).

Total subsidy, as in the Canadian case, is exceptional. More common is the partial subsidy, which still does harm because it commonizes the costs over the whole body politic while privatizing the profitable aspects among a much smaller group. A subsidy is sometimes symbolized as a "CC-PP Game" (Commonize Costs–Privatize Profits).[1] Each beneficiary of this game receives a (relatively) large personal benefit from it, while each victim—each taxpayer—suffers only a small personal loss. The beneficiaries of the game are like the New Jersey stevedores cited earlier: their consciences don't hurt them because they are sure that the pain of loss to any one individual is trifling in comparison with the pleasures of those who profit from the subsidy. All insurance against loss is one wing of a CC-PP Game, the existence of which discourages the insuring agent from trying too hard to rout out minor thievery.

Many or most of the members of a community receive both gains and losses from the game. The WorldWatch Institute estimated that the world's catch of fish sold for $70 billion in 1989, whereas catching the fish cost $124 billion. The difference ($54 billion) was subsidized by a variety of taxes.[2] Many individuals benefit from cheaper fish while losing in the taxes they pay for the subsidy. Their posterity will also lose because the subsidy encourages practices that will eventually deplete the fish populations.

The average citizen, driven by mixed motives, finds that "the native hue of resolution," as Hamlet put it, "is sicklied o'er with the pale cast of thought, and enterprises of great pith and moment . . . lose the name of action."[3] More bluntly: inaction becomes the order of the day.

Subsidies have great survival power. Decades ago, the U.S. government started paying for the forest roads and other infrastructures sought by the lumber interests. Taxpayers are still subsidizing the destruction of the forests. Long ago our government established the practice of leasing public land to cattle ranchers at much less than the amount charged by private landowners; we are still doing that also. In the tobacco-growing states we have, for generations, been making it possible for farmers to grow tobacco profitably; during the same period, we have been spending even more money combatting the adverse effects of tobacco on the health of tens of millions of taxpayers. It is as though one hand knows not what the other hand is doing.

Actually, the blame must be assigned to the assymetry of benefits and costs in the CC-PP Game. Commonized costs are easy to ignore when they are small per capita, but who wants to give up large privatized profits? Playing this game is a serious seduction in all forms of government. (It's too bad that posterity cannot be present to vote on it.)

▲ ▼ ▲ ▼ ▲

We are now ready to see why the so-called population problem is so resistant to solution. In today's dominant cultures, two rights are asserted simultaneously:

1. *The right to life.* In practice, we support the solemn pronouncement of the United Nations that *every man, woman, and child has the inalienable right to be free from hunger and malnutrition.*

2. *Reproductive right. Every woman has the right—perhaps with the*

agreement of her mate(s)—to determine how many children she shall produce. (Nature is, of course, a sleeping partner in this agreement; she sometimes has other plans.)

In the plays of ancient Greece the playwright sometimes became so entangled with the plot that no rational resolution seemed possible. Rather than admit defeat, the author took the coward's way out: he invented an agent of the gods who, by supernatural means, engineered a happy ending. The agent was called a *deus ex machina*— "god from a gadget" I guess we'd say now.

A modern scientist, using Einstein's technique for solving the difficult problem of species survival, would ask, "How would God manage it?" He or she would soon conclude that the potential increase of a species must always be exponential: that is, like money put out at compound interest. But if the resources *practically* available to the species are finite, the species will soon eat itself out of house and home and die. That won't do. So Einstein's God must supply a deus ex machina (or several) to every species to keep it from committing suicide. That is where the "blessings of Tertullian" come in, and the greatest of these is disease.

The infectivity of pathogenic agents is subject to a scale effect: the denser the susceptible population, *the larger the proportion* of the population that will die of disease. Now human beings are interfering with the ancient balance. Man the inventor and discoverer—*Homo faber*— has seriously injured, and may eventually kill, the Tertullian deus ex machina.

The preeminent question for modern humankind is this: can we assemble a new deus ex machina that will ensure the survival of our species? This new miracle worker must be internal to humanity itself—inside our heads, nestled within the commitments we make to society. It is clear that reproduction will have to have its rights trimmed. Many minds will have to contribute to the planning of the commitments that must modify the unqualified reproductive right if humanity is to survive. *This* is the population problem of our time—not collecting and playing with a plethora of ambivalent data.

I think we human beings can manage this—*provided that* we

rid ourselves of certain assumptions that have served us well in the past but that cannot do so without the spice of a blessing à la Tertullian. I'm assuming that no new overwhelming disease will turn up. Those who have lived through the arrival of AIDS may find this a hazardous assumption, but I am enough of an optimist to think it is not. (If I am wrong in this assumption, a new disease will function as a new deus ex machina, and we're back to living in the Tertullian mode.)

In the exposition that remains, I will take what is called the optimistic position, discussing the pessimistic conclusion to which, paradoxically, it leads. This "problem of problems" will demand humanity's most imaginative attention for a long time to come.

▲ ▼ ▲ ▼ ▲

The foundations for the new efforts must include the insights of two men cited in chapter 12: Bertrand Russell and Goldwin Smith. At the level of simple commonsense knowledge of human nature, Russell reminded us that the irreducible cantankerousness of human beings would ensure that One World, if achieved, would soon break up. We would be left once more with the problem of war, which One Worlders thought their program would solve. Perhaps the most we can hope for is a partial taming of war. (Converting these soothing words into specific directives is another major problem. We are encouraged to note that World War III, so confidently predicted in the near future immediately after the atomic bomb was invented, has not yet arrived, though more than half a century has passed. *Knock on wood!*)

Russell's argument seems a counsel of despair to many people, but almost a century before Russell, a less well known Englishman, Goldwin Smith, said that there is an important way in which a fragmented world is safer than a world under a single sovereign power: "If all mankind were one state, with one set of customs, one literature, one code of laws, *and this state became corrupted*, what remedy . . . would there be?" (emphasis added)

We can recast Smith's argument to involve our old friend, the Man from Mars. Imagine that we have thoroughly studied all 180+ sovereign nations in the world, noting the great variety of

customs and commitments among them. At this point our oblig-
ing Martian says: "I guess you didn't know that I am a genie. I
offer you One Worlders the chance of selecting a single national pattern of government, with the sole restriction that your own
nation cannot be your choice of model, because you cannot
judge its qualities objectively. Which one of the 180 nations of
the world would you be content to accept as the universal government?"

Perhaps in the late 1920s the most popular choice would have
been the young and vigorous USSR. Would we all be living as
Soviet citizens now? Probably not: the USSR lasted only for the
biblical life span of 70 years. Then it collapsed in chaos. Between
1930 and 1990 the population of the USSR more than doubled,
growing from about 135 million to 290 million. Did that growth
make this nation more stable or less? History gave the answer.
The disastrous collapse of the USSR was no fluke. As Goldwin
Smith foresaw, a large-scale unity would soon be followed by "a
convulsion which would rend the frame of society to pieces, and
deeply injure the moral life which society is designed to guard."
The chaos of the collapsed USSR will present the world with
many, many problems for a long time to come.

Even if there is such a thing as a "best government," we dare
not assume that we know what it is. Not knowing, it is wise to
accept a multisystem world, with the variants physically isolated
by effective immigration barriers and tariff walls so that there
can be meaningful competition of one culture with another. For
information, however, there should be no barriers; safe trans-
mission of information can take place via printed matter, electronic interchanges, and strictly limited personal visits.

*Inter*nationally, multiculturalism would be the order of the
day. It would help observing nations to measure one culture
against another, thus making possible a multiple experiment in
political arrangements. International multiculturalism would,
for a long time, be the default position; but in the interests of
peace, every nation would be well advised not to accept ficti-
tious "universal human rights" as an excuse for interfering in
the affairs of other nations. The history of foreign interven-
tions by the United States is one of repeated and embarrassing
failures brought about by our thinking that good intentions

necessarily produce good results in utterly different cultures.

*Intra*national standards of behavior must be intolerant of multiculturalism: this attitude is the default position. Driving a car is an element of culture, and citizens cannot safely be allowed to drive on either the left or right side of the road as they wish. This extreme example illuminates the peril in all proposals of multiculturalism within a nation.

Population characteristics also derive from elements of culture—pronatalism, for instance. In this area, as in many others, complete freedom is intolerable. If individuals can move freely into and out of nations, then the Extended Gresham's law applies: *low standards of living drive out high standards* as the world moves toward universal poverty under the banner "Need creates right!" The right of immigration will be denied; acceptance of immigrants must *always* be in the interest of the receiving country. Because posterity and the future should be involved in the calculus of immigration standards, it is difficult to specify the best level of immigration. (At the present time, the United States is taking in—legally and illegally—something like 2 million immigrants per year. Faced with severe standards of admission, would more than 100 persons per year be admissible? Such a question will be labeled *bigoted* by the ethnofugalists, who cannot honestly face the problems posed by radically different cultural ideals.)

The movement of physical wealth is equally intolerant of complete freedom. Substantial tariffs must be the rule. Because of the complexity of industrial evolution, arriving at standards will always be difficult, but the default position should be: *no hidden domestic subsidies for foreign products.* The most powerful enemies of restrictions in tariffs are local enterprisers seeking to maximize their own wealth at the expense of their local compatriots.

▲ ▼ ▲ ▼ ▲

Most of what has been said above will prove abhorrent to large numbers of Americans (as well as to many other modern people). *What shall we strive for—growth or stability? Should we try to maximize possessions, and rates of utilizing them, or minimize them?* It is interesting to note that Malthus, two decades after publishing his celebrated essay, raised this problem in his textbook on political economy. What sort of gift is acceptable to someone who is concerned about the continued well-being of a particular community?

> It is not easy to conceive a more disastrous present—one more likely to plunge the human race in irrecoverable misery, than an unlimited facility of producing food in a limited space. A benevolent Creator then, knowing the wants and necessities of his creatures, under the laws to which he has subjected them, could not, in mercy, have furnished the whole of the necessaries of life in the same plenty as air and water.[4]

For the past two centuries, most commentators on population problems have spoken as if the biblical commandment to be fruitful and multiply means *forever*. But does God give a prize for the maximum number of people? Such a God cannot be Einstein's God. If such is now the God of Christian sects, they are putting their money on the wrong horse.

Setting aside the question of the real infinity of the universe, it is clear that civilization has now advanced to a level where the rate of increase in technology by far exceeds the rate at which we dare create new demands on the available environment as a result of increases in population or in its demands made on the environment by rising standards of living. If the optimization of living conditions is the standard of judgment, then we must say, from here on out, that population growth must be minimized. Some say that the growth rate should even be negative for a while (but that argument can be put on the back burner at present).

What images should fuel the ambitions of our youth? In the 19th century, Jules Verne's fictional exploration of the worlds of submarines and spaceships called forth great efforts among the youth who were intoxicated by these dreams. Space continued to be an important intoxicant of the 20th century.

What about the 21st century? At this point, the conscientious adviser of the young faces a dilemma. For some decades to come, growth-oriented youth will no doubt succeed in life better than the small minority that have their doubts about continued growth.

On the other hand, a hint of another kind of advice is found in an experience of one of the talented scientists of the second half of the 20th century: F. Sherwood Rowland of the University of California at Irvine. Rowland and Mario Molina (from Mexico City) discovered that the very useful and supposedly harmless substances called *chlorofluorocarbons* (*CFCs*) were destroying the ozone in the earth's upper atmosphere, letting in more ultraviolet light. The total effects on human affairs are hard to predict, but already there are legitimate worries. Here is a form of technological progress that may, in fact, ultimately be disastrous for *Homo sapiens* (as well as for many other species).

Embarking on their professional careers, beginning graduate students go where the action is. Without doubt, there was real action at Irvine. But for 14 years after Rowland's basic discovery, no graduate students outside the University of California system applied to study and work with him. Only after 1988, when DuPont gave him a large research contract, did Rowland start to get a substantial number of doctoral candidates from outside California. The trend increased after he received the Nobel Prize in chemistry in 1995.[5]

Why the long delay in the acceptance of this meritorious research by the upcoming generation of students? Rowland himself has no doubt as to the cause: he was discovering facts that undermined an important secular element of faith in our time, namely, that important discoveries in science *always* improve the human situation in the natural world. A century earlier, scientists discovered that certain bacteria turned the essentially useless molecular nitrogen in the atmosphere into nitrogen compounds that could enrich the soil and increase our food supply. But Rowland had discovered that the massive use of CFCs (and the creation of masses of financial wealth) had a delayed effect that might, at worst, turn our planet into another Venus (where the surface temperature is around 800 degrees Fahrenheit because of long-ago contamination of its atmosphere).

Improvement is an optimistic idea; research students expect to benefit financially from optimistic discoveries. Disimprovement is pessimistic; canny students doubt that they will benefit from being part of a team that makes pessimistic discoveries. Legions struggle to get through the door of optimism; the doorway of pessimism is uncrowded. So what should the elders of a discipline say to their neophytes?

At the age of 70, and full of honors, Rowland says to the young worker who is hoping to make a reputation in a crowded field: "Don't look under the light. Go out into the darkness."

People who ask, "What is the maximum population that the earth can support?" are just spinning their wheels. Surely God doesn't give a prize for the maximum population. If He is concerned at all with our well-being, it is hard for a thinking human being to believe that He would work only in the light. The big difficulties, worthy of the attention of God worshippers, are in the dark now, in the area that, after many centuries of being neglected, promises to make human life more enjoyable. But investigating the problems of minimizing rates, or optimizing sizes, or reconciling conflicting ends can be fiendishly difficult. But in those dark corners lie the greatest possibilities of discovering reforms capable of improving the human condition.

At the level of pure wisdom, we need more knowledge about the consequences of population-linked cultural habits. Knowledge alone will not move nations: astonishing and unforeseen events will be required for humanity's education. But the details of history cannot be foreseen, only some overriding characteristics of human reactions. Inertia is one of them. Born of the human life cycle is this generalization: "'From time immemorial' means, in practice, 'for three generations.' If an assertion was true in my grandfather's time, in my father's time, and now in mine, I don't see how I can doubt it."

In that sense, perpetual growth in a world of infinite and available material resources has been treated as an immemorial truth. Painful experiences will be required to banish this illusion from the intellectual armamentarium of humanity's leaders. Our ostriches will have to have their heads yanked out of the comforting sands of illusion.

In the interim, we can take heart from the history of the opening up of space in the 20th century. The young who, in the early days, fantasized about traveling in spaceships were (we can now see) dreaming against the tide. Eventually the tide turned; at the most basic economic level, it was realized that the flame of landing on distant planets was not worth the candle of monstrous taxes. If the dream of perpetual growth is now near its end, then it is time to explore the possibilities of living in a non-growing but sustainable world, a world in which temperance in global ambitions is a virtue. The forced world of skilled publicity agents for growth-intoxicated industrialists will be seen to be less promising than the subtle world of Rowland's dark area. When this last becomes the world of the young, we will make genuine progress in optimizing the quality of life on earth.

Notes

The writer of a trade book is often well advised to omit notes and references. A scholarly work is apt to be loaded with references that are seldom consulted; this is part of the "one-upmanship" of the scholarly enterprise. The present work is designed to be something in between. When the reference is to a whole book, the author and title appearing in the main text are enough to lead readers to the volume, provided that they have access to a good municipal or university library. Trenchant quotations, the origins of which can be found in standard collections (e.g., Bartlett's *Quotations*), are not identified further in this text. The principal citations will be of documents that might not otherwise be known to the reader.

Chapter 1

1. See J. A. Bierens de Haan, "On the Ostrich Which Puts Its Head in the Sand in Case of Danger, or: The History of a Legend," *Ardea*, 32:11–24, 1943.
2. The first edition (1798) was correctly called *An Essay on the Principle of Population*. This, the most accessible of Malthus's writings, has been republished many times. Later editions (beginning in 1803) were treatises rather than essays. The University of Michigan's reprint (1959) of the first edition has a long and insightful foreword

by Kenneth E. Boulding. Also highly recommended is the volume edited by Philip Appleman (New York: Norton, 1976), which contains selections from several of the editions together with many supporting and divergent views ranging from those of William Godwin (1797) to Pope Paul VI (1968).

3. Credit for inventing the Man from Mars, an aid to objective thought, has been given to two different Frenchmen, but I have been unable to verify either claim. Many intellectuals, from the Greeks to Francis Bacon, have tried to step back from their personal prejudices, even from their own cultures. In 1721, in his *Lettres Persannes*, Montesquieu invented an imaginary traveler from Persia who sent home critical descriptions of French life and traditions that would no doubt have gotten the author into hot water had they been presented as straightforward reports by Montesquieu himself. The Man from Mars, a spiritual descendant of Montesquieu's Persian commentator, should be in the reader's mind as he or she tries to look objectively at our most treasured (and least questioned) presuppositions.

4. Herman E. Daly, *Beyond Growth: The Economics of Sustainable Development* (Boston: Beacon Press, 1996). The introduction provides a brief and powerful presentation of ecological economics by Daly, one of its founders.

5. Sagoff's review of my book is found in *Trends in Ecology and Evolution*, 9(12):498-499, 1994.

6. Hans Spemann, *Embryonic Development and Induction* (New Haven, CT: Yale University Press, 1938), p. 367.

Chapter 2

1. Tertullian, *De Anima*, 3rd century A.D. It can be found in Garrett Hardin, *Population, Evolution and Birth Control*, 2nd ed. (San Francisco: Freeman, 1969), p. 18. See also Bertrand Russell on the innate imperialism of life in *An Outline of Philosophy* (London: Unwin Paperbacks, 1979), p. 22.

2. Daniel Bell, quoted in Kenneth D. Wilson, ed., *Prospects for Growth: Changing Expectations for the Future* (New York: Praeger, 1977), pp. 18-19.

Chapter 3

1. Herbert Spencer, *The Principles of Psychology*, 2nd ed. (London: Norgate, 1860), p. 396. For the parody by P. G. Tait, see *Nature*, 23:80-82, 1880. This exchange of comments is recorded in Daniel C. Dennett,

Darwin's Dangerous Idea (New York: Simon & Schuster, 1995), p. 393n.

2. George Orwell, *A Collection of Essays* (San Diego: Harcourt Brace Jovanovich, 1981), p. 229.

3. Paul Demeny, "Population and the Invisible Hand," *Demography*, 23(4):473–478, 1986 (quote on p. 473).

Chapter 4

1. On Catholic rights, see August Bernhard Hasler, *How the Pope Became Infallible: Pius and the Politics of Persuasion* (Garden City, NY: Doubleday, 1981), p. 41. See also Germain G. Grisez, "On *Humanae Vitae*," *National Catholic Reporter*, 4(40):8, 1968.

2. Jeremy Bentham, *Anarchical Fallacies* (French publication in 1816; English publication in 1843). See Jack Parsons, *Population vs. Liberty* (London: Pemberton, 1971), p. 131.

3. Lacking, like most modern scientists, a proper education in the classical languages of Latin and Greek, I find the motto *Nullius in verba* somewhat puzzling. Kenneth Brown, a well-educated philosopher friend, informed me that any of several free translations should be acceptable:
   1. "I am obliged to swear to the words of no master."
   2. "I belong to no school of philosophy."
   3. "On no one's authority."
   4. "On no one's say-so."

   I have chosen the third. For a further discussion see Henry Allen Moe, "Tercentenary of the Royal Society," *Science*, 132:1817, 1960.

4. Cyril Bailey, *Epicurus, The Extant Remains* (Oxford: Clarendon Press, 1925), p. 21.

5. See Garrett Hardin, "Paramount Positions in Ecological Economics," in Robert Costanza, ed., *Ecological Economics: The Science and Management of Sustainability* (New York: Columbia University Press, 1991), pp. 47–57. The abstract for this paper had to be submitted before the symposium at which it was presented. It was in the abstract that I introduced the term *default position*. By the time I produced the paper itself, I had decided that *paramount position* was a better term. Ever since then, I have felt that my first judgment was the sounder—that *default* is more suited to the tentative nature of science than is *paramount*. *L'uomo e mobile!*

6. Alice Calaprice, ed., *The Quotable Einstein* (Princeton, NJ: Princeton University Press, 1996), p. 76.

Chapter 5

1. Harold J. Barnett and Chandler Morse, *Scarcity and Growth* (Baltimore: Johns Hopkins University Press, 1963), p. 12.
2. Peter T. Bauer, *Equality, and the Third World and Economic Delusions* (Cambridge, MA: Harvard University Press, 1981), p. 206.
3. George Gilder, *Wealth and Poverty* (New York: Basic Books, 1981).
4. Paul W. McCracken, "A Way Out of the World's Slump," *Wall Street Journal*, 17 September 1975, p. 24.
5. Charles Perrings, *Economy and Environment* (Cambridge: Cambridge University Press, 1987), p. 129.
6. Julian Simon, "The State of Humanity: Steadily Improving," *Cato Policy Report*, 17(5):131, 1995. A remarkable fact about the Cato Institute (before which Simon frequently spoke) is that, though it is populated largely by people conventionally educated in the economic sciences, ecological economics seems intolerable to them. The remarks by Hugh Iltis on Simon's sale of indulgences is from a personal communication of 29 March 1989. To properly appreciate the unintended humor in Simon's public utterances, one needs to be familiar with a poem by Arthur Guiterman in his collection *Lyric Laughter* (New York: Dutton, 1939). The poem, "The Vizier's Apology" (pp. 173–174), is quite appropriate to a discussion of the way Julian Simon reacted to criticism. When Simon corrected his monumental error of a billion years by changing it to a million years, he unconsciously revealed that he had not the foggiest notion of the power of exponential growth. A "mere" million years would yet carry the function off the map of conceivable human experience. Considering the central importance of money at interest in economic thinking, it is shocking to find a presumed economist making such an error—or an institute filled with economists publishing it. Thus did Simon and his associates prove Guiterman's point: "an excuse might be worse than the crime."
7. Kenneth Arrow et al., "Economic Growth, Carrying Capacity, and the Environment," *Science*, 268:520–521, 1995.
8. Albert A. Bartlett, "The Exponential Function, XI: The New Flat Earth Society," *Physics Teacher*, 34:342–343, 1996. In addition to a series of papers on exponential growth, Professor Bartlett has given a lecture on the subject to more than a thousand university audiences. It is nominally the same lecture, superbly polished but varying slightly in detail every time it is repeated.
9. Ken Alder, "The Perpetual Search for Perpetual Motion," *American Heritage of Invention and Technology*, 2:58–63, 1986. After 1918, the Patent Office refused to accept applications for perpetual motion machines.

In 1930, all old applications were burned. Such are the consequences when Epicurus's assumptions are taken as the default position.

10. The term *longage* was introduced in the poem "Carrying Capacity." See Hardin, *Stalking the Wild Taboo*, 3rd ed. (Petoskey, MI: Social Contract Press, 1996), pp. 316–318.

11. Robert K. Merton, *The Unanticipated Consequences of Purposive Action* (New York: Free Press, 1976). This essay, originally published in 1936, is a fine statement of the ecological outlook on human attempts at reform.

Chapter 6

1. For a lay reaction by Mrs. Seagraves, see the *Washington Post*, 4 March 1981, p. A-7.

2. The reaction of W. H. Auden to evolution can be found in the preface to Loren Eiseley, *The Star Thrower* (New York: New York Times Books, 1978).

3. See J. Koskimies, "The Life of the Swift," *Annales Academiae Scientiarum Fennicae, Series A, IV. Biologica*, 12:1–151, 1950.

4. Richard D. Alexander, "The Evolution of Genitalia and Mating Behavior in Crickets (Gryllidae) and other Orthoptera," *Miscellaneous Publications of the Museum of Zoology, University of Michigan*, 133:1–62, 1967.

5. Hardin, "An Ecological View of Ethics," in James B. Miller and Kenneth E. McCall, eds., *The Church and Contemporary Cosmology* (Pittsburgh: Carnegie Mellon University Press, 1990), p. 345.

Chapter 7

1. The literature on Charles Darwin and evolution is, of course, immense. The following publications provide good introductions: Arthur O. Lovejoy, *The Great Chain of Being* (Cambridge, MA: Harvard University Press, 1936), in which the historical origins of the idea of evolution are described; Neal C. Gillespie, *Charles Darwin and the Problem of Creation* (Chicago: University of Chicago Press, 1979), p. 15 of which records the "ungodding" of the universe; Morse Peckham, "Darwin and Darwinisticism," *Victorian Studies*, 3:19–40 (1959), which includes a record of Darwin's reluctance to use the word *evolution*.

2. Philip P. Wiener and Aaron Noland, eds., *Roots of Scientific Thought* (New York: Basic Books, 1957), p. 529.

3. *Life and Letters of the Reverend Adam Sedgwick* (Cambridge: Cambridge University Press, 1890), vol. 2, p. 440. Report of a lecture given in

1868, the 83rd year of Sedgwick's life.

4. Dorothy L. Sayers, *The Nine Tailors* (New York: Harbrace, 1962), p. 58.

5. *Science*, 215:34, 1982.

6. Gottfried Wilhelm Leibniz in Charles Vereker, *Eighteenth-Century Optimism* (Liverpool: Liverpool University Press, 1967), p. 17.

7. Banesh Hoffman, *Albert Einstein, Creator and Rebel* (New York: New American Library, 1972), p. 18.

8. Charles Kingsley, in Francis Darwin, ed., *Letters of Charles Darwin* (New York: Appleton, 1898), vol. 2, p. 82.

9. The July 1995 issue of *Scientific American* (p. 112) gives a provocative account of tactics used by American fundamentalists to expunge even the mention of Darwinism from public school curricula. Thus is censorship solidified in the legal support of ostrichism.

Chapter 8

1. Hermann Heinrich Gossens, 1854. See Herman E. Daly and John B. Cobb, *For the Common Good* (Boston: Beacon Press, 1989), p. 89.

2. F. Y. Edgeworth, *Mathematical Psychics* (London: F. K. Paul, 1881), p. 16.

3. C. S. Lewis, 1942. *The Screwtape Letters* (London: Fontana Books, 1955), pp. 134–135.

Chapter 9

1. William Ophuls and Stephen Boyen, Jr., *Ecology and the Politics of Scarcity Revisited* (New York: W. H. Freeman, 1992), p. 200–201.

2. William Paley, *Natural Theology*, 1802 (London: Gilbert & Rivington, 1885), p. 520.

3. Claude-Adrien Helvétius, *De l'Esprit*, 1758. Taken from the *Encyclopaedia Brittanica*, 1974, vol. 6, p. 891.

4. On their altruism, see Donald R. Kraybill, *The Riddle of Amish Culture* (Baltimore: Johns Hopkins University Press, 1989), p. 92.

5. Will and Ariel Durant, *Rousseau and Revolution* (New York: Simon & Schuster, 1967), p. 282.

6. John Locke, *An Essay Concerning Human Understanding*, 1689 (Cleveland: Meridian Books, 1964), pp. 223–225.

Chapter 10

1. George's much reprinted treatise proposed to discourage making profits from speculating in unimproved property. Speculators fought his tax proposals vigorously. Price inflation continued.

2. Robert K. Merton, "The Self-Fulfilling Prophecy," 1948, in Merton, *On Social Structure and Science* (Chicago: University of Chicago Press, 1996), pp. 183–201.
3. Jorge G. Casteñada. *Los Angeles Times,* 12 February 1993, p. B7.
4. Personal communication, 11 January 1994.

Chapter 11

1. Mortimer J. Adler, as reported in John E. Rankin, quoting from the *Cleveland Plain Dealer,* 29 October 1945; in the *Congressional Record,* 1 November 1945, p. A4662.
2. Barbara Ward, *The Rich Nations and the Poor Nations* (New York: Norton, 1962), pp. 152–153.
3. Marshall McLuhan, *Understanding Media: The Extensions of Man* (New York: Penguin, 1964), p. 20.
4. Kenneth J. Arrow and 22 others, *Mount Carmel Declaration of Technology and Moral Responsibility* (Haifa, Israel: Technion-Israel Institute of Technology, 25 December 1975).

Chapter 12

1. Bertrand Russell, *Authority and the Individual* (London: Allen & Unwin, 1948), pp. 16–17.
2. E. B. White, *The Wild Flag* (Boston: Houghton Mifflin, 1946), pp. xii, 10–11.
3. Goldwin Smith, 1860, as reported in Jacob Viner, *The Role of Providence in the Social Order* (Philadelphia: American Philosophical Society, 1972), p. 49.

Chapter 13

1. Malthus, *An Essay on the Principle of Population,* 2nd ed. (London: J. Johnson, 1803), p. 503.
2. This verse was first published in *Perspectives in Biology and Medicine,* 9:225, 1966.
3. The basic insight was known to Darwin and was rediscovered several times during the 20th century. The first full explication of it appears in my essay "The Competitive Exclusion Principle," *Science,* 131:1292–1297, 1960. For a more accessible treatment of it, see "The Cybernetics of Competition" in my collection *Stalking the Wild Taboo.*

Chapter 14

1. Alexis de Tocqueville, *Democracy in America* (New York: Vintage Books, 1945), vol. 2, p. 102.
2. Helmut Schoeck, *Envy* (New York: Harcourt, Brace & World, 1966).
3. Isaiah Berlin, *Concepts and Categories* (London: Hogarth Press, 1978), p. 84.
4. John Stuart Mill in Andre Beteille, *The Idea of Natural Equality* (Delhi: Oxford University Press, 1987), p. 157c.
5. In *Science*, 204:1285, 1979. Also in *Los Angeles Times*, 18 August 1979 (part 1, p. 26), and 21 December 1979 (part 1, p. 3).
6. Alfred North Whitehead, *The Aims of Education* (New York: Mentor, 1949), p. 19.

Chapter 15

1. W. G. Sumner, *Folkways* (New York: Dover Pub., 1959), p. 13.
2. E. E. Evans-Pritchard, *Social Anthropology* (London: Cohen & West, 1951), p. 127.
3. C. P. Snow, "The Moral Un-Certainty of Science," *Science*, 133:256–259, 1961 (quote on p. 258).
4. Richard Bernstein, *Dictatorship of Virtue* (New York: Knopf, 1994), pp. 7–8.
5. "The Cultural Defense in Criminal Law," *Harvard Law Review*, 99(6):1293–1311, 1986.
6. Fred Siegel, "Is Archie Bunker Fit to Rule?" *Telos*, 69:9–31, 1986, p. 26.
7. Merton, "The Unanticipated Consequences of Social Action," 1936, reprinted in Merton, *On Social Structure and Science* (Chicago: University of Chicago Press, 1996), pp. 173–182.
8. Nathan Glazer, *Affirmative Discrimination: Ethnic Inequality and Public Policy* (New York: Basic Books, 1975).
9. On the ordeal of Alan Bakke, see Victor A. Walsh and Thomas E. Wood, *Bakke and Beyond: A Study of Racial and Gender Preferences in California Public Higher Education* (Berkeley: California of Scholars, 1730 Martin Luther King Jr. Way, 1996).

Chapter 16

1. Malthus, *Principles of Political Economy*, 1820 (Fairfield, NJ: Kelley, 1986), p. 208.
2. Joel E. Cohen, *How Many People Can the Earth Support?* (New York: Norton, 1995).
3. J. H. Fremlin, "How Many People Can the World Support?" *New Scientist*, 415:285–287, 1964.

Chapter 17

1. Cicero, *De Officiis* (On Duties), ed. Harry G. Edinger (New York: Bobbs-Merrill, 1974), p. 163.
2. John M. Culbertson, *The Dangers of Free Trade* (Madison, WI: 21st Century Press, 1985).
3. Roy Beck, *The Case against Immigration* (New York: Norton, 1996). This is a balanced and insightful treatment of the problem in an American setting.
4. Interdisciplinary insight reveals that even more general than Gresham's law is the idea of selection. This can be stated as a tautology: "The fittest is that which survives." The tautology is not, however, trivial: personifying nature, we can say that "nature invests in success." This is a true statistical summary of experience. When we find that the "success" that nature exalts is not what we want, we must constrict liberty to achieve our ends. Thus, we enact laws against counterfeiting; without these, reliable commerce is impossible. Similarly, enforcible borders protect us against the universalism of poverty through the cross-boundary movements of either human bodies or salable human artifacts. In all these cases, deliberate discrimination is essential to a defense against the evils of promiscuity. Achieving control of human population numbers, once Tertullian's "blessings" are made nugatory, will require discrimination in the dispensation of the right to reproduce—even as the earlier creation of a trustworthy money system required some loss of human freedom.

Chapter 18

1. For an extended discussion of the properties and powers of the CC-PP game, see Garrett Hardin, *Filters against Folly* (New York: Viking, 1985), chap. 10.
2. Periodical publications of the WorldWatch Institute should be followed closely.
3. William Shakespeare, *Hamlet,* Act 3, Scene 1, line 56, 1601.
4. Malthus, *Principles of Political Economy,* 1820, (Fairfield, NJ: Kelley, 1986), pp. 208-209.
5. Bill Burton, "Clean-up Hitter," *Chicago,* 89(6):13-15, 1997, an account of the difficulties that faced Rowland as he pursued challenging research with pessimistic implications (in the view of compulsive optimists).

Index